IFP Energies nouvelles Publications

▶ Jean-François NAUROY
IFP Energies nouvelles

GEOMECHANICS APPLIED TO THE PETROLEUM INDUSTRY

2011

Editions TECHNIP 5 av. de la république, 75011 PARIS, FRANCE

FROM THE SAME PUBLISHER

- Heavy Crude Oils
  From Geology to Upgrading. An Overview
  A.Y. HUC
- CO_2 Capture
  Technologies to Reduce Greenhouse Gas Emissions
  F. LECOMTE, P. BROUTIN, E. LEBAS
- Corrosion and Degradation of Metallic Materials
  Understanding of the Phenomena and Applications in Petroleum and Process Industries
  F. ROPITAL
- Multiphase Production
  Pipeline Transport, Pumping and Metering
  J. FALCIMAIGNE, S. DECARRE
- A Geoscientist's Guide to Petrophysics
  B. ZINSZNER, F.M. PERRIN
- Acido-Basic Catalysis (2 vols.)
  Application to Refining and Petrochemistry
  C. MARCILLY
- Petroleum Microbiology (2 vols.)
  Concepts. Environmental Implications. Industrial Applications
  J.P. VANDECASTEELE
- Physico-Chemical Analysis of Industrial Catalysts
  A Practical Guide to Characterisation
  J. LYNCH
- Chemical Reactors
  From Design to Operation
  P. TRAMBOUZE, J.P. EUZEN
- Petrochemical Processes (2 vols.)
  Technical and Economic Characteristics
  A. CHAUVEL, G. LEFEBVRE
- The Technology of Catalytic Oxidations (2 vols.)
  P. ARPENTINIER, F. CAVANI, F. TRIFIRO
- Marine Oil Spills and Soils Contaminated by Hydrocarbons
  C. BOCARD

Printed in France

ISBN 978-2-7108-0932-6

Preface

According to great organisations involved in energy prospective studies as IEA (International Energy Agency) and DOE (US Department of Energy), world energy demand will still grow in the coming years and fossil energies will contribute in the range of 60 up to 80% to the energy supply, oil and gas contributing to about 50% of the energy supply. This has two major implications. On one hand, oil and gas will have to be produced from more and more difficult fields; on the other hand, more CO_2 will be emitted and will have to be captured and stored.

Both fluid production and CO_2 storage involve fluid flow within porous rocks. This induces pressure changes, saturation changes, temperature changes with water injection or thermal recovery, even fluid interaction with the rock for CO_2 injection. All these phenomena have an impact on stresses and strains, or on the mechanical behaviour of rocks.

Rock mechanics in the oil and gas industry was traditionally used for drilling, wellbore stability and fracturing. Reservoir application came with the observed consequences of subsidence and the associated reservoir compaction. It should be stressed here that to satisfy the increase in energy demand, fluids will be produced from complex environments where the mechanical behaviour of the rock will be more and more important, for instance for wellbore stability or faults behaviour.

A new area is emerging with CO_2 management and more specifically CO_2 sequestration. Massive storage will occur by injecting in deep saline aquifers. Along with the description of the reservoir behaviour, the knowledge of the overburden behaviour from a mechanical standpoint will be essential to assess the storage integrity.

This book covers all these aspects and aims at providing a large range of potential users with a simple approach of a broad field of knowledge. The reader will enjoy the tables of reference values for the needed parameters such as wave velocity, elastic properties given for both static and dynamic properties. Furthermore, the Figures are easy to understand and illustrate very accurately the theoretical development. The book is application oriented, thus closed forms, or empirical formulae are provided to give first estimates of the problem. For instance, the reader will find indicators of drilling efficiency, closed forms for hydraulic or thermal fracturing or for subsidence prediction. The practical consideration is always nearby. Hence, sections are devoted to lab determination of properties, *in situ* determination of stresses, sand production prevention and finally geomechanical monitoring of reservoirs.

While rock mechanics is of importance in many fields of application such as drilling, production or reservoir engineering, it is still thought of as a complex field restricted to a few experts. I hope and even guess that, through this book, any engineer involved in those fields will understand the basics and the practical applications of rock mechanics.

Maurice Boutéca
Director
Resources Business Unit
IFP Energies nouvelles

Acknowledgements

I first want to thank Pascal Longuemare, who initiated this book, and all our friends and colleagues from *IFP Energies nouvelles* who defined its content, provided scientific and technical material to feed it, and wrote or reviewed parts of it.

I address special thanks to authors who wrote one or more parts of the book. It has been a great pleasure to work together:

Elisabeth Bemer	Chapters 1.2, 1.3, 3.1
Marc Mainguy	Chapter 3.2
Claude Putot	Chapter 2.1
Sandrine Vidal-Gilbert	Chapters 1.1, 3.3

The authors would like to thank Florence Adjemian, Axelle Baroni, Frédérique Fournier, Pascal Longuemare and Bernard Zinszner for their partial to total rereading and their constructive suggestions.

They express their deep gratitude to Atef Onaisi (Total) and Jean-Paul Sarda (expert at *IFP Energies nouvelles*) for their thorough review and precious comments.

They express their thanks to Maurice Boutéca for having accepted to preface the book.

Lastly, the authors are grateful to Patrick Boisserpe and Mireille Darthenay for unfailing patience throughout the project and their precious support, through their active and fruitful proofreading of the manuscript. Our thanks also go to Dominique Allinquant for the drawings.

Jean-François Nauroy
Editor

Acknowledgements

Table of Contents

Chapter 2
GEOMECHANICS, DRILLING AND PRODUCTION

Chapter 3
GEOMECHANICS AND RESERVOIR

List of Authors

Elisabeth Bemer
Engineer from the *École Nationale des Ponts et Chaussées*
PhD in Structures and Materials from the *École Nationale des Ponts et Chaussées*
Research Engineer at IFP Energies nouvelles (since 2000) in the field of Geomechanics, Laboratory Manager
Petrophysics Department, Reservoir Engineering Division
(IFP Energies nouvelles, 92852 Rueil-Malmaison Cedex, elisabeth.bemer@ifpen.fr)

Marc Mainguy
Applied Mathematics Engineer from *Université Paris XIII, Institut Galilée*
PhD in Structures and Materials from the *École Nationale des Ponts et Chaussées*
Postdoctoral researcher (in 1999) at MIT, Department of Civil & Environmental Engineering
Research Engineer at IFP Energies nouvelles (from 2000 to 2005) in the field of Geomechanics
Geomechanics Engineer at TOTAL, Drilling & Wells, Well Productivity Engineering Department
(TOTAL, CSTJF, avenue Larribau, 64018 Pau Cedex, France, marc.mainguy@total.com)

Jean-François Nauroy
Geophysicist Engineer from IFP School
PhD in Soil Mechanics from the Paris VI University *(Pierre et Marie Curie)*
Research Engineer at IFP Energies nouvelles (since 1977) in the field of Soil and Rocks Mechanics
Petrophysics Department, Reservoir Engineering Division
(IFP Energies nouvelles, 92852 Rueil-Malmaison Cedex, j-francois.nauroy@ifpen.fr)

Claude Putot
Engineer from the *École Nationale Supérieure des Mines de Paris*
PhD in Fracture Mechanics from the Paris VI University *(Pierre et Marie Curie)*
Former Principal Research Engineer with IFP Energies nouvelles in the field of Structural Analysis and Rock Mechanics

Sandrine Vidal-Gilbert
Geophysicist Engineer from *Institut de Physique du Globe de Strasbourg* and from *IFP School*
PhD in Geosciences from Paris VI University *(Pierre et Marie Curie)*
Research Scientist at IFP Energies nouvelles (from 2000 to 2006) in the field of Geomechanics
Senior Researcher in CO2CRC at the University of Adelaide (from 2006 to 2008)
Project Leader at TOTAL for Enhanced Coal Bed Methane (ECBM) and Underground Coal Gasification (UCG) R&D projects
(TOTAL, Gas&Power, Place Jean Millier, 92078 Paris La Défense, sandrine.vidal-gilbert@total.com)

Main Symbols

ρ density
Φ Lagrangian porosity
ϕ Eulerian porosity
ε void ratio
ρ_s solid matrix density
ρ_{fl} fluid density

k permeability
μ fluid viscosity
η dynamic viscosity

$\underline{\underline{\varepsilon}}$ elastic strain tensor

$\underline{\underline{\sigma}}$ stress tensor
$\underline{\underline{\sigma}}'$ effective stress tensor

Pp pore pressure
T temperature

ν_d drained Poisson's ratio
ν_u undrained Poisson's ratio

E Young's modulus
G shear modulus

b Biot's coefficient

K_d drained bulk modulus (bulk modulus of the skeleton)
K_u undrained bulk modulus
K_s bulk modulus of the solid matrix
K_{fl} bulk modulus of the fluid

α_d skeleton thermal linear coefficient of expansion
α_u thermal linear coefficient of expansion
α_Φ pore thermal linear coefficient of expansion
α_{fl} pore fluid thermal linear coefficient of expansion

V_p compression wave velocity
V_s shear wave velocity

λ_T thermal conductivity
λ plastic compressibility coefficient
κ swelling coefficient

g free enthalpy
S, s entropy
H, h specific enthalpy
ψ free energy
Gs potential

Introduction

At all times during the exploitation of an oil field, the engineers strive to optimise drilling times, reduce uncertainties and production costs, and optimise the use of available data.

Geomechanics, which includes all thermohydromechanical phenomena, plays an important role in every operation involved in the exploitation of hydrocarbons, from drilling to production and right up to the time the wells are abandoned. Pressure changes in the reservoir modify the *in situ* stresses and cause strains not only to the reservoir but also to the entire sedimentary column. In turn, these stress modifications and strains change the stability parameters of the walls of the wells to be drilled and the flow properties of the fluids.

The main purpose of this book is to provide engineers involved in oil field development and petroleum engineering students, whether drillers or producers, the information they need about the techniques and approaches currently used in geomechanics.

The book is divided into three broad sections.

The first section provides information on rock mechanics, describes the characterisation of rocks in the laboratory and the modelling of their mechanical behaviour. This chapter also describes how geomechanics determine the stresses applied to the material in place and why they need to model the reservoirs and the surrounding material.

The second section deals with the role of geomechanics in the optimisation of drilling (drillability and wellbore stability) and production. Sand production, which brings numerous undesirable problems, can be predicted and in some cases controlled. Hydraulic fracturing can be encouraged to stimulate production or on the contrary avoided so as not to reactivate a fault, in case of CO_2 sequestration.

The third section deals with the coupling between the production of fluids in the reservoir and the mechanical strains of the complete oil structure. The environmental aspects of this interaction are discussed with respect to subsidence of the surface layers, CO_2 sequestration and well abandonment. Geomechanical monitoring of reservoirs may prove essential in some cases to optimise production or check cap rock integrity.

Warning! In rock mechanics the compressive stresses are positive, whereas in continuous media mechanics, stresses are negative. The latter convention is adopted in the numerical models. Each chapter will indicate whether the positive or negative compressive stresses are being used.

1

Elements of Rock Mechanics Fundamentals

1.1 BASIC NOTIONS

This chapter provides a general introduction to rock mechanics dedicated to petroleum applications. Some simplifications have been required to respect the objective set. For more details, refer to more specific books on petroleum geology and fundamental rock mechanics.

1.1.1 Rock characteristics

1.1.1.1 Minerals and rocks

The earth's crust accessible to geological observation consists of rocks. Almost all rocks are aggregates of minerals, forming grains, visible under the naked eye or the microscope. Minerals are natural chemical species, generally consisting of crystalline solids, characterised by a three-dimensional periodic arrangement of atoms or ions. More than 4,000 minerals are known. Silica and silicates are the most abundant [Guillemot, 1986].

The physical, chemical and mechanical properties of rocks depend on the physical, chemical and mechanical properties of the constituent minerals, the volume fraction of each mineral and the distribution and arrangement of the minerals with respect to each other. The microstructure of this assembly is rarely homogeneous: arrangement of minerals, mineralogical variability, distribution of grain sizes, spatial distribution of the grains, degree of fracturing, etc. The rock heterogeneity depends on the scale of observation, especially since its physical properties depend on the scale of measurement: centrimetric for laboratory samples to metric, or even kilometric, for blocks *in situ*.

1.1.1.2 Rock classification

There are three broad classes of rock: magmatic, sedimentary and metamorphic.

1.1.1.2.1 Magmatic rocks

Magmatic rocks are mainly silicates. They result from the solidification of a magma (liquid rock, possibly containing crystals in suspension). When the magma cooled very slowly deep within the lithosphere, the grains (crystals) are well developed: this is the case of plutonic (or intrusive) rocks, such as granite. When the magma cooled quickly after irrupting on the surface or at the bottom of the oceans, the grains are generally very small, apart from the occasional macrocrystals: volcanic (or extrusive) rocks, such as basalt.

1.1.1.2.2 Sedimentary rocks

Although magmatic rocks represent about 95% of the earth's crust (by volume), 75% of the area of the continents and virtually all the ocean floors are covered with sedimentary rocks resulting from the accumulation and compaction of mineral debris – in this case we speak of detrital rock (degradation of other rocks), organic debris (plant or animal remains, fossils) –, or chemical precipitation.

When the sediments are deposited in lacustrine and marine environments, they are soft and generally water saturated. They are converted into coherent rocks under the effect of diagenesis, a term referring to several phenomena that can occur to varying extents depending on the deposition conditions and the type of sediment concerned (compaction, cementation, recrystallisation). Sedimentary rocks are brought to the surface during the large tectonic movements which take place in the earth's crust. When the ocean or continental plates collide, the sediments are folded, deformed and uplifted to form valleys and high mountain belts.

During these tectonic movements, the rock may have been subjected locally to very high pressures and temperatures.

Sedimentary rocks are classified into four groups:

- *Sandstones* are detrital sedimentary rocks composed of grains of sand (quartz or silica) bound together by a siliceous, calcareous or *ferruginous* cement which determines the strength of the sandstone. Quartz and feldspars are the main components. Sandstone elements belong to the granulometric class of arenites (2 mm to 0.05 mm).
- *Shales* are defined as detrital rocks of fine grain size distribution (grain diameter less than 2 μm). Phyllites (chlorite, montmorillonite, muscovite and kaolinite) are the major component minerals, but quartz, feldspar and calcite are also present. Shales have very complex properties.
- *Carbonate rocks* are composed mainly of the minerals calcite, aragonite and dolomite, of origin either organic (marine organisms) or inorganic (chemical precipitation). Calcite is the dominant mineral.
- *Evaporites:* these rocks are composed of elements precipitated from water. As the water evaporates, the species in solution become more and more concentrated, up to oversaturation then precipitation of the components. The main minerals precipitated are gypsum, anhydrite and halite.

1.1.1.2.3 Metamorphic rocks

Metamorphic rocks are formed from mineralogical transformations of pre-existing rocks (sedimentary or magmatic) without fusion, mainly through recrystallisations due to temperature and pressure increases.

1.1.1.3 Sedimentary rock and porous medium

Sedimentary rocks are porous media composed of a deformable skeleton and interstitial voids, which are extremely important for oil companies. In terms of geomechanics, these voids account for the major differences in behaviour observed between the compact solids (e.g. metals) and the porous media, complex assemblies where the presence of fluids, even in small quantities, makes the rock even more complex. For further information on the mechanics of porous media, refer to the books by Coussy (1995) and Bourbié *et al.* (1986).

The total porosity of a given rock is the ratio of the non-solid volume (pore volume) to the total volume

$$\phi = \frac{V_p}{V} = \frac{V - V_s}{V} \tag{1.1.1}$$

where V_p is the porous volume,
V_s the solid volume,
V the total volume.

This porosity is also called Eulerian porosity, to distinguish it from Lagrangian porosity used in modelling $\Phi = \frac{V_p}{V_0}$, *with respect to the initial total volume* V_0.

There are numerous ways of measuring porosity but the measurement result may depend on the method used [Bourbié *et al.*, 1986].

This petrophysical quantity provides no information on the pore dimensions, distribution or degree of connection.

The porosity of a sediment varies during burial. The sediment changes progressively from a quasi-liquid state to a compact solid state. The initial intergranular porosity may be very high, 70% to 80% for a clay sediment. During burial of the sediment, the confining pressure and the temperature increase. Various diagenetic processes (mechanical, chemical and biochemical) change the porosity by consolidation and compaction of the rock, dissolution and recrystallisation of the minerals, etc.

Extensive theoretical and experimental studies have been conducted on the flow of fluids through the porous media. The permeability of a porous medium is a property which reflects the ability of the rock to allow a fluid to flow through its pores. The absolute permeability is an intrinsic characteristic of the material, independent of the fluid. This permeability k was defined by Darcy, by demonstrating that the discharge rate Q through a porous medium of cross-sectional area A was proportional to the pressure drop ΔP, divided by the dynamic viscosity μ of the saturating fluid and the length L over which the pressure drop is taking place:

$$Q = -A\frac{k}{\mu}\frac{\Delta P}{L} \tag{1.1.2}$$

The negative sign is needed since fluids flow from high to low pressure.

When the following units are used: Q in $m^3.s^{-1}$, A in m^2, $\Delta P/L$ in $Pa.m^{-1}$ and μ in Poises, the permeability k is expressed in m^2 and in Darcy, with 1 Darcy $= 0.97.10^{-12}$ m^2. The Darcy, denoted D, is widely used in petroleum engineering and petrophysics. The m^2 is used more by material physicists.

The hydraulic conductivity K, or permeability coefficient, is defined as the ratio of permeability to fluid viscosity:

$$K = \frac{k}{\mu} \tag{1.1.3}$$

K is expressed in $m.s^{-1}$. So, while the permeability k is a characteristic of the material that is highly dependent on the porosity, the hydraulic conductivity K is characteristic of the conditions of flow in a given material for a given fluid. To a certain extent, the hydraulic conductivity K takes into account the physical interactions between the fluid and the rock.

The permeability depends on the structure of the porous medium. Two rocks with the same porosity may have quite different permeabilities. When the porous space is filled by several fluids, the capillary effects act in addition to the dynamic head losses and, in this case, permeabilities relative to the fluid considered can be determined.

1.1.1.4 Properties of heterogeneous media

Application of the macroscopic equations of mechanics to porous media assumes that these media are continuous, in other words that physical values (porosity, permeability, saturation, etc.) can be defined at each point as functions differentiable at the point considered. At microscopic scale, however, a point of the rock is either in the solid or in the pores, thereby forming a discontinuity (if we take the variable "porosity" as an example). The notion of "Representative Elementary Volume" (REV) is related to the need to describe the properties of a porous medium at a scale where the structural discontinuities have no influence on these properties. Consequently, the dimensions of the elementary volumes required to include the effect of the discontinuities may vary in particular for a given medium, depending on the parameter considered.

When a macroscopic constitutive equation is defined for a given medium, these various minimum homogenisation volumes must be taken into account. The mechanical experiments will only be relevant "macroscopically" if the sample is much larger than the minimum homogenisation volume of the parameter considered. When dealing with cracked media, this volume is sometimes highly underestimated.

1.1.2 Stresses and strains

Analysis of stresses and strains is essential whenever studying rock mechanics. Since these notions have been studied extensively in numerous books [e.g. Jaeger and Cook, 1969], only a few details will be given here.

1.1.2.1 Stresses

For the simple case of an axially loaded body, e.g., a cylindrical bar subjected to tension or compression by a force passing through its centre, the stress σ, or intensity of the distribution of internal forces, can be obtained by dividing the total tensile or compressive force amplitude F by the cross-sectional area A over which it is acting. In this case the stress is a scalar that represents an average stress σ_{avg} over the area, meaning that the stress in the cross section is uniformly distributed. We therefore obtain:

$$\sigma_{avg} = \frac{F}{A}$$

In general, however, the stress is not uniformly distributed over a cross section of a material body, and consequently the stress at a point is different from the average stress over the entire area. The local stress at a specific point in the body is defined as being the limiting value of $\Delta F / \Delta A$ when ΔA tends to zero.

If the force is not applied perpendicular to the cross-section A, the stress can be decomposed into a normal component called the *normal stress* and a tangential component called the *shear stress*.

- **Stress tensor**

To provide a complete description of the stress state at a given point, we must be able to identify the stresses relative to the surfaces in three orthogonal directions. There are 9 components of the stress at a point, given by the tensor expression:

$$\underline{\underline{\sigma}} = \begin{pmatrix} \sigma_x & \tau_{xy} & \tau_{xz} \\ \tau_{yx} & \sigma_y & \tau_{yz} \\ \tau_{zx} & \tau_{zy} & \sigma_z \end{pmatrix}$$

- **Invariants**

Invariants remain unchanged during a change of coordinate system:

The mean normal stress is equal to the trace of the stress tensor matrix:

$$\sigma_{mean} = \left(\sigma_x + \sigma_y + \sigma_z\right) / 3 \qquad (1.1.4)$$

The other invariants commonly used are:

$$I_2 = -\left(\sigma_x\sigma_y + \sigma_y\sigma_z + \sigma_z\sigma_x\right) + \tau_{xy}^2 + \tau_{yz}^2 + \tau_{zx}^2$$
$$I_3 = \sigma_x\sigma_y\sigma_z + 2\tau_{xy}\tau_{yz}\tau_{zx} - \sigma_x\tau_{yz}^2 - \sigma_y\tau_{zx}^2 - \sigma_z\tau_{xy}^2 \qquad (1.1.5)$$

- **Deviatoric stress**

The average stress defined previously expresses a uniform compression or extension. Distortions are expressed by the *deviatoric stress* tensor. Deviatoric stresses can be obtained by subtracting the mean normal stress components from the stress tensor.

$$\begin{pmatrix} s_x & s_{xy} & s_{xz} \\ s_{yx} & s_y & s_{yz} \\ s_{zx} & s_{zy} & s_z \end{pmatrix} = \begin{pmatrix} \sigma_x - \sigma_{mean} & \tau_{xy} & \tau_{xz} \\ \tau_{yx} & \sigma_y - \sigma_{mean} & \tau_{yz} \\ \tau_{zx} & \tau_{zy} & \sigma_z - \sigma_{mean} \end{pmatrix}$$

- **Principal stresses**

At any point of a body subjected to a system of forces, three planes can be defined on which there are only normal stresses and no shear stresses. These planes are called the *principal planes* and the stresses the *principal stresses*. In this coordinate system, the stress tensor has a simple expression.

$$\begin{bmatrix} \sigma_1 & 0 & 0 \\ 0 & \sigma_2 & 0 \\ 0 & 0 & \sigma_3 \end{bmatrix}$$

The three principal stresses are known as the major, intermediate and minor principal stresses, in descending order of magnitude.

- **Mohr circle**

In 2D, knowing the principal stresses σ_1 and σ_2 ($\sigma_1 > \sigma_2$), the normal and shear stresses can be calculated in any other plane inclined at an angle θ in the *xy* plane (Figure 1.1.1).

Solving the equilibrium equations (sum of forces equal to zero) implies:

$$\sigma_\theta = \frac{1}{2}\left(\sigma_1 + \sigma_2\right) + \frac{1}{2}\left(\sigma_1 - \sigma_2\right)cos2\theta$$
$$\tau_\theta = -\frac{1}{2}\left(\sigma_1 - \sigma_2\right)sin2\theta \qquad (1.1.6)$$

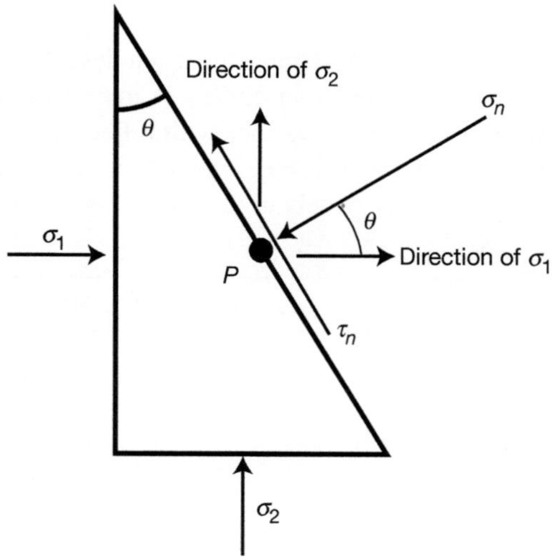

Figure 1.1.1

State of stress at a point.

The stress state (σ_θ, τ_θ) in a plane at an angle θ to the horizontal describes a circle in the plane (σ, τ), called Mohr's circle. Figure 1.1.2 shows a semi-circle of radius $\frac{1}{2}(\sigma_1 - \sigma_2)$ and centre at $\frac{1}{2}(\sigma_1 + \sigma_2)$.

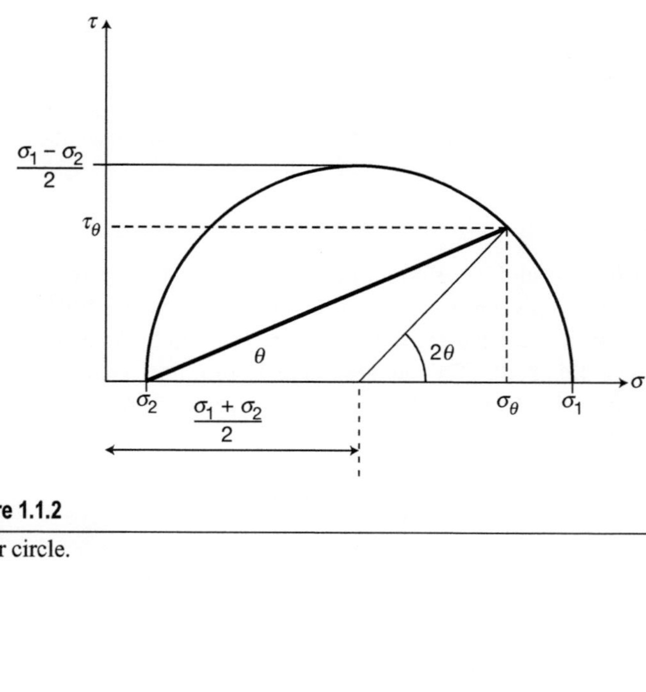

Figure 1.1.2

Mohr circle.

- ***p-q* diagram and stress path**

Successive stress states applied to a rock element are often described in geomechanics. One way of doing this is to draw a series of Mohr circles. A more satisfactory way is to plot a series of points and connect these points together with a line or a curve. These points are called stress paths [Lambe and Whitman, 1969]. Due to the easy generalisation offered, the p-q diagram (Figure 1.1.3) is currently widely used, where p is the mean principal stress and q is the deviatoric stress:

$$q = \sqrt{\frac{(\sigma_1 - \sigma_2)^2 + (\sigma_2 - \sigma_3)^2 + (\sigma_1 - \sigma_3)^2}{2}} \tag{1.1.7}$$

$$p = \frac{\sigma_1 + \sigma_2 + \sigma_3}{3}$$

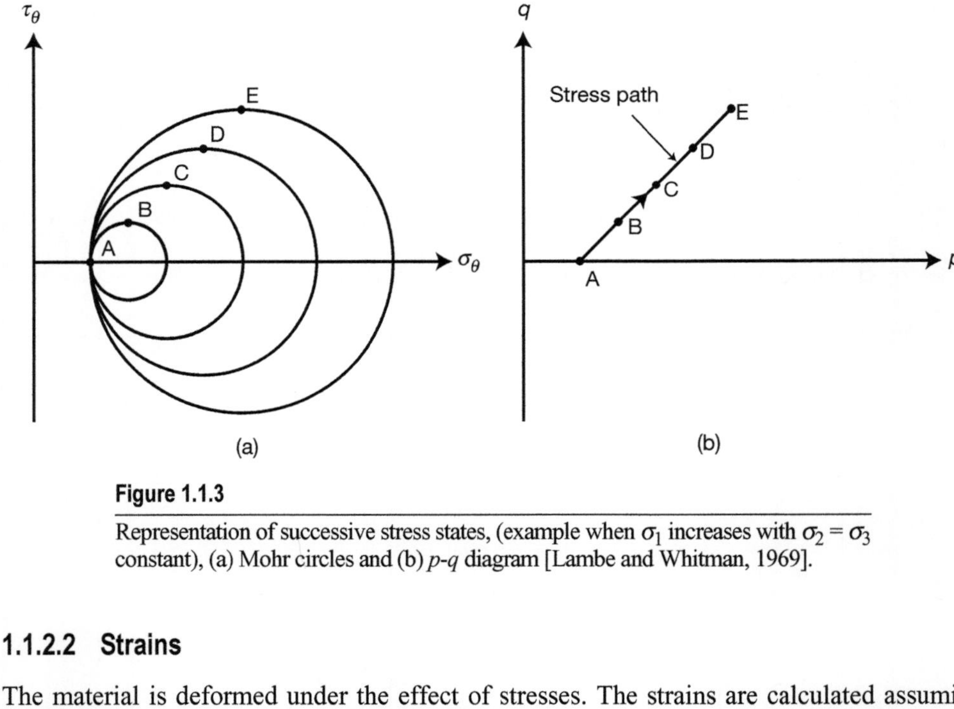

Figure 1.1.3

Representation of successive stress states, (example when σ_1 increases with $\sigma_2 = \sigma_3$ constant), (a) Mohr circles and (b) *p-q* diagram [Lambe and Whitman, 1969].

1.1.2.2 Strains

The material is deformed under the effect of stresses. The strains are calculated assuming there has been a change between two states: the initial state and the final state. The strain of a body can be measured by calculating the change in length of a line on the body or the angle between two lines on this body. The change in length of a line, an elongation or a contraction, can be denoted as δl. The strain ε, a dimensionless ratio, is defined as the ratio of elongation with respect to the original length:

$$\varepsilon = \frac{\delta l}{l_0} = \frac{l - l_0}{l_0} \tag{1.1.8}$$

with

ε strain in the direction measured,
l_0 initial length of the line,
l current length of the line.

The local strain in the given direction can be expressed as:

$$\varepsilon = lim \frac{\delta l}{l}$$

when l tends to zero.

Angular strain γ is defined by the amount of deformation Δl perpendicular to a given line rather than parallel to it. The angular deformation is also called the shear strain (measured in radians):

$$\frac{\Delta l}{l} = tg\gamma = \gamma$$

- **Strain tensor and invariants**

Assuming infinitesimal strains, if a point undergoes a displacement by a vector $\underline{u}$, the mathematical expression of the strains is given by:

$$\varepsilon_{ij} = \frac{1}{2}\left(\frac{\partial u_i}{\partial x_j} + \frac{\partial u_j}{\partial x_i}\right) \tag{1.1.9}$$

The strain tensor is written:

$$\underline{\underline{\varepsilon}} = \begin{pmatrix} \varepsilon_{xx} & \varepsilon_{xy} & \varepsilon_{xz} \\ \varepsilon_{yx} & \varepsilon_{yy} & \varepsilon_{yz} \\ \varepsilon_{zx} & \varepsilon_{zy} & \varepsilon_{zz} \end{pmatrix}$$

The tensor trace $\varepsilon_v = \varepsilon_{xx} + \varepsilon_{yy} + \varepsilon_{zz}$ expresses the *volume strain*. This value, independent of the coordinate system chosen, is an invariant. Other strain invariants can be defined:

$$\begin{aligned} J_2 &= -\left(\varepsilon_x\varepsilon_y + \varepsilon_y\varepsilon_z + \varepsilon_z\varepsilon_x\right) + \varepsilon_{xy}^2 + \varepsilon_{yz}^2 + \varepsilon_{zx}^2 \\ J_3 &= \varepsilon_x\varepsilon_y\varepsilon_z + 2\varepsilon_{xy}\varepsilon_{yz}\varepsilon_{zx} - \varepsilon_x\varepsilon_{yz}^2 - \varepsilon_y\varepsilon_{zx}^2 - \varepsilon_z\varepsilon_{xy}^2 \end{aligned} \tag{1.1.10}$$

As for the stresses, the strain tensor can be written with respect to its principal planes.

1.1.2.3 Effective stress

The pore pressure plays a fundamental role studied through the notion of effective stress, used to take into account the twofold effect of the stress (σ_{ij}) and the pore pressure (p_p) via an expression, which is a linear combination of these two variables:

$$\sigma'_{ij} = \sigma_{ij} - \alpha\, p_p\, \delta_{ij} \quad \text{or} \quad \underline{\underline{\sigma}}' = \underline{\underline{\sigma}} - \alpha\, p_p \underline{\underline{1}} \tag{1.1.11}$$

where δ_{ij} is the Kronecker tensor.

Since this concept is applied to problems involving distinct mechanisms – constitutive equations, failure criterion and rock characteristics – the corresponding expressions of the effective stress do not necessarily coincide. When the coefficient α is equal to 1, we speak of Terzaghi's effective stress, named after the inventor of the concept for soil mechanics.

For more in-depth information on this topic, refer to [Guéguen and Boutéca, 1999].

1.1.2.4 State Boundary Surface (SBS)

Using the critical state concept, soil mechanics scientists have developed the notion of stress path. Stress paths plotted in the 3-dimensional space p'-q-v (p': mean effective stress, q: deviatoric stress, v: volume variation) determine a unique surface known as the State Boundary Surface (SBS) (Figure 1.1.4). This surface separates the area of the possible

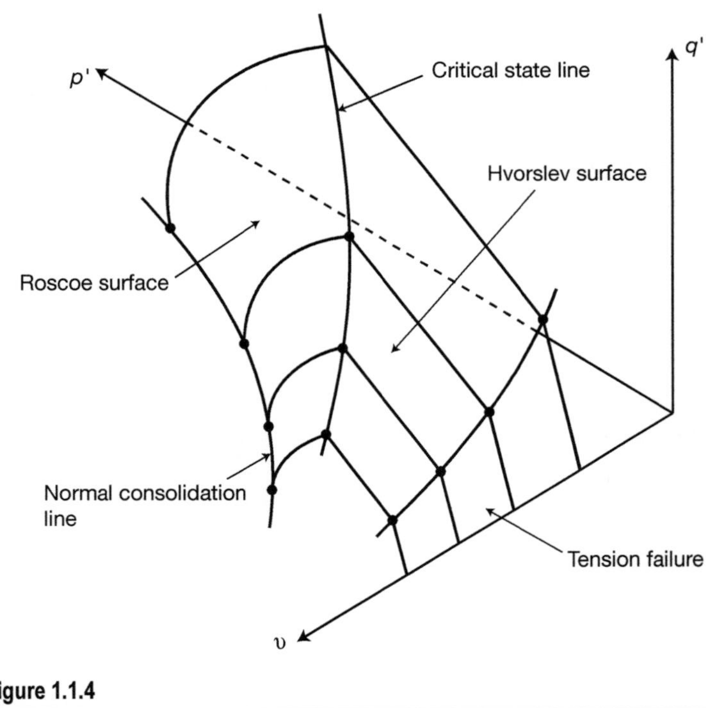

Figure 1.1.4

State boundary surface [Schofield and Wroth, 1968].

"stress-volume" states from the area where "stress-volume" states cannot exist. The states presented by the parameters (p', q, v) always lie inside or on the SBS. The response of soils inside the SBS is purely elastic (reversible but not necessarily linear) and becomes plastic when the soil state reaches the state boundary surface. This notion has been extended to rock mechanics. The SBS is also called yield surface.

1.1.2.5 Elastic properties

1.1.2.5.1 Static moduli of elasticity

The theory of elasticity describes situations where there are linear relations between the stresses applied and the resulting strains. We consider a cylindrical sample of length l and cross-sectional area A to which a force F is applied at its two ends. The stress applied to the sample is $\sigma_z = \frac{F}{A}$. The sample will undergo an axial strain of $\varepsilon_z = \Delta l / l$ and two radial strains ε_x and ε_y ($\varepsilon_x = \varepsilon_y$ for a cylindrical sample). The strains in this type of experiment are related to the stresses by the following linear relation:

$$\varepsilon_z = \frac{\sigma_z}{E} \text{ relation known as Hooke's Law.} \tag{1.1.12}$$

E, called the *modulus of elasticity* or *Young's modulus*, characterises the stiffness of the sample, in other words its resistance to an axial stress. The radial deformability can also be characterised using the *Poisson's ratio*:

$$v = -\frac{\varepsilon_x}{\varepsilon_z} \tag{1.1.13}$$

For isotropic materials, whose response is independent of the direction of the stresses applied, the relations between the stresses and strains are given by the generalised *Hooke's Law*:

$$\varepsilon_{ij} = \frac{1}{E}\left[(1+v)\sigma_{ij} - v\delta_{ij}\sigma_{\alpha\alpha}\right]$$

or (1.1.14)

$$\sigma_{ij} = \lambda\delta_{ij}\varepsilon_{\alpha\alpha} + 2G\varepsilon_{ij}$$

where λ and G, the *Lamé parameters,* are also elastic moduli. G is called the *shear modulus,* since it characterises the resistance of the sample to shear. The shear modulus is the ratio of shear stress to shear strain.

The *bulk modulus K* defines the ratio of an isotropic stress to a volume strain. In an isotropic stress state $\sigma_x = \sigma_y = \sigma_z$, the other components of the stress tensor are zero: $K = \frac{\sigma_p}{\varepsilon_v}$.

K is the sample resistance to an isotropic compression. The reciprocal $1/K$, called compressibility, is written C_p. The various elastic constants and their definitions are given in

Table 1.1.1. Note that for porous media, we must introduce K_d and K_u, bulk moduli, respectively in drained and undrained conditions, E_d and E_u, drained and undrained Young's moduli, and v_d and v_u, drained and undrained Poisson's ratios.

Table 1.1.1 Definitions of elastic constants [Guéguen and Placiauskas, 1992].

	K	E	v	G
Experiment	Isotropic compression σ_p	Uniaxial compression σ_z	Uniaxial compression σ_z	Shear σ_{xy}
Definition	$K = \frac{\sigma_p}{\varepsilon_v}$	$E = \frac{\sigma_z}{\varepsilon_z}$	$v = -\frac{\varepsilon_x}{\varepsilon_z}$	$G = \frac{\sigma_{xy}}{\varepsilon_{xy}}$

The relations between the various elastic constants K, E, G, λ and v are listed in Table 1.1.2.

Table 1.1.2 Relations between the various elastic constants of an isotropic material [Birch, 1961].

K	E	v	G	λ
$\lambda + \frac{2}{3}G$	$G\frac{3\lambda + 2G}{\lambda + G}$	$\frac{\lambda}{2(\lambda + G)}$	—	—
—	$9K\frac{K-\lambda}{3K-\lambda}$	$\frac{\lambda}{3K-\lambda}$	$\frac{3}{2}(K-\lambda)$	—
—	$\frac{9KG}{3K+G}$	$\frac{3K-2G}{2(3K+G)}$	—	$K - \frac{2}{3}G$
$\frac{EG}{3(3G-E)}$	—	$\frac{E}{2G} - 1$	—	$G\frac{E-2G}{(3G-E)}$
—	—	$\frac{3K-E}{6K}$	$\frac{3KE}{9K-E}$	$3K\frac{3K-E}{9K-E}$
$\lambda\frac{1+v}{3v}$	$\lambda\frac{(1+v)(1-2v)}{v}$	—	$\lambda\frac{1-2v}{2v}$	—
$G\frac{2(1+v)}{3(1-2v)}$	$2G(1+v)$	—	—	$G\frac{2v}{1-2v}$
—	$3K(1-2v)$	—	$3K\frac{1-2v}{2+2v}$	$3K\frac{v}{1+v}$
$\frac{E}{3(1-2v)}$	—	—	$\frac{E}{2+2v}$	$\frac{Ev}{(1+v)(1-2v)}$

1.1.2.6 Dynamic moduli of elasticity

The moduli of elasticity listed above are determined by experiments on rock samples subjected to tension – compression (static strain). They are called the static moduli of elasticity. These moduli of elasticity can also be determined by measuring the velocities of propagation of seismic (or acoustic) waves (dynamic strain). In this case, they are called the dynamic moduli of elasticity.

The compression V_p and shear V_s wave velocities are given by the following expressions:

$$V_P = \sqrt{\frac{K_{dyn} + \frac{4}{3} G_{dyn}}{\rho}} \tag{1.1.15}$$

$$V_S = \sqrt{\frac{G_{dyn}}{\rho}} \tag{1.1.16}$$

When the *P* and *S* wave velocities are available, the following dynamic moduli of elasticity can be determined:

$$G_{dyn} = \rho V_S^2 \quad \text{and} \quad K_{dyn} = \rho V_P^2 - \frac{4}{3}\rho V_S^2 \tag{1.1.17}$$

The dynamic values of Young's modulus and the Poisson's ratio are given respectively by:

$$E_{dyn} = \frac{\rho V_S^2 \left(3V_P^2 - 4V_S^2\right)}{\left(V_P^2 - V_S^2\right)} \quad \text{and} \quad \nu_{dyn} = \frac{V_P^2 - 2V_S^2}{2\left(V_P^2 - V_S^2\right)} \tag{1.1.18}$$

The static moduli of rocks are generally lower than the dynamic moduli, due to the amplitudes of the strains considered. In oil exploration, seismic data, Vertical Seismic Profile (VSP) data and P- and S-wave logging data provide measurements of propagation velocities for rocks around and between the wells, both in the reservoir and in the surrounding strata. These measurements are not recorded at the same frequency (Table 1.1.3), but data of velocity dispersion as a function of frequency is interesting for petroacousticians.

Table 1.1.3 Frequency of acoustic waves.

	Acquisition frequency
Exploration seismic	100-300 Hz
VSP	100-300 Hz
Sonic logging	100 kHz
Measurement on samples in laboratory	0.5 to 1 MHz

Seismic velocities are sensitive to lithology, porosity, fluid content, temperature and stress field. Orders of magnitudes of the velocities V_p and V_s for some rocks are provided for information in Table 1.1.4.

Table 1.1.4 Order of magnitude of P and S wave velocities and density for different rock types [Lavergne, 1986].

Rock type	P wave velocity (m/s)	S wave velocity (m/s)	Density (g/cm^3)
Scree, organic soil	300-700	100-300	1.7-2.4
Dry sands	400-1,200	100-500	1.5-1.7
Wet sands	1,500-2,000	400-600	1.9-2.1
Saturated clays	1,100-2,500	200-800	2.0-2.4
Marls	2,000-3,000	750-1,500	2.1-2.6
Saturated argillaceous-sandstone series	1,500-2,200	500-750	2.1-2.4
Porous and saturated sandstones	2,000-3,500	800-1,800	2.1-2.4
Limestones	3,500-6,000	2,000-3,300	2.4-2.7
Chalk	2,300-2,600	1,100-1,300	1.8-2.3
Salt	4,500-5,500	2,500-3,100	2.1-2.3
Anhydrite	4,000-5,500	2,200-3,100	2.9-3.0
Dolomite	3,500-6,500	1,900-3,600	2.5-2.9
Granite	4,500-6,000	2,500-3,300	2.5-2.7
Basalt	5,000-6,000	2,800-3,400	2.7-3.1

1.1.2.7 Anisotropy

If the elastic response of the material depends on its orientation for a given stress field, the material is said to be *anisotropic*. In this configuration, the moduli of elasticity vary depending on the direction in which they are evaluated.

Anisotropy is defined as heterogeneities at a smaller scale than that of the volume observed. At large scale, the development of geological strata (foliations, lineations, etc.) results in a certain degree of anisotropy, while at small scale this anisotropy exists due to the presence of the anisotropic minerals forming the rocks.

For anisotropic medium, a generalised form of Hooke's law is written:

$$\sigma_{ij} = C_{ijkl}\varepsilon_{kl} \quad \text{or} \quad \underline{\underline{\sigma}} = \underline{\underline{\underline{\underline{C}}}}\,\underline{\underline{\varepsilon}} \tag{1.1.19}$$

$\underline{\underline{\underline{\underline{C}}}}$, elastic stiffness tensor, is a 9x9 tensor. The symmetry of the stresses and strains reduces the number of independent components to 36. In addition, the existence of a unique energy potential reduces the number of independent components to 21, the maximum number of elastic constants that a medium can possess.

Most sedimentary rocks (sandstone, mudstone, shale, etc.) exhibit anisotropic characteristics in terms of strength and deformational properties (differences between elastic moduli measured parallel and perpendicular to bedding), and are generally considered to be transversely isotropic materials, symmetric about an axis normal to a plane of isotropy. The number of independent elastic constants is reduced to 5 (2 Young's moduli, 2 Poisson's ratios and one shear modulus). Some rocks may have three distinct directions of anisotropy. These materials are orthotropic and have 9 independent elastic constants (3 Young's moduli, 3 Poisson's ratios and 3 shear moduli).

1.2 GEOMECHANICAL CHARACTERISATION OF ROCKS IN THE LABORATORY

Apart from its structural and mineralogical variability at microscopic scale, at macroscopic scale a rock mass may exhibit pronounced stratification, cracks, faults, etc. Generally therefore, it is not homogeneous. The fundamental hydromechanical properties of a rock are considered to be those of a sample big enough to contain a large number of component grains, but small enough to exclude any major structural discontinuity. More precisely, it is recommended that the diameter of a sample should be more than 10 times the diameter of the largest rock grain [Kovari *et al.*, 1983]. Samples of a few centimetres generally correspond to the mesoscopic scale so defined and offer the advantage of being easily tested in the laboratory.

The aim of laboratory tests is to characterise the hydromechanical properties of a rock using relatively simple experiments. The properties determined are based on a certain number of assumptions (homogeneous stress and strain states in the sample, isotropic behaviour, etc.). It is also essential to ensure that they are actually representative of the behaviour of the rock, not of the experimental device. Relations between stresses and strains can generally be deduced from the measurements taken. These relations are then extrapolated to deal with more complex situations at a scale different from that of the tests, this scale change requiring additional assumptions.

1.2.1 Laboratory tests

In situ, a rock is subjected to a triaxial stress state and to the pressure exerted by the interstitial fluid(s). The stresses are generally assumed to be oriented in the vertical direction and two orthogonal horizontal directions, the vertical stress being determined by the overburden (see Chapter 1.4).

The time scale of the strains may vary considerably depending on the problem considered: for instance, the strain rates associated with progressive depletion of a reservoir are much slower than those encountered when drilling a well. The strength of a rock generally increases with strain rate. In the laboratory, strain rates between 10^{-6} and 10^{-1} s^{-1} are generally applied. Within this interval (not necessarily representative of the problem studied), the properties of most rocks can be considered as globally independent of the strain rate.

However, with very poorly permeable rock, such as shale, the strain rate chosen must be low enough to prevent excess pore pressure generation.

The various types of test used to study the hydromechanical characteristics of a rock, in particular its failure properties, are mainly characterised by the specific stress states imposed and the accessible properties. The associated cost and ease of implementation must also be taken into consideration when selecting the tests to be conducted during a characterisation campaign.

This chapter outlines the main types of test on rock and the properties that can be determined. We will then focus on the compressive behaviour of a rock, using the results of the main rock mechanics test: the conventional triaxial test.

This document does not discuss hollow cylinder tests that can be used to study well wall behaviour [Fjaer *et al.*, 1992] [Jaeger and Cook, 1979].

1.2.2 Tests without interstitial fluid

1.2.2.1 Uniaxial compression test

The uniaxial compression test (also called unconfined compression test) is one of the oldest and easiest to implement. It consists in compressing a generally cylindrical rock sample between the two platens of a load frame. An increasing axial compressive force is applied to the sample, resulting in its gradual shortening and eventual failure (Figure 1.2.1). The axial force applied and the associated axial strain are measured. The radial strain can also be measured. The axial stress is equal to the force measured divided by the initial cross-section of the sample. With a highly deformable rock, it may be worthwhile evaluating the current cross-section of the sample from the measurement of radial strain and calculating a "true" axial stress.

The Unconfined Compressive Strength (*UCS*) of the rock (value of the axial stress peak associated with the failure) and estimations of its Young's modulus and its Poisson's ratio are deduced from this test. Assuming linear elastic behaviour, these parameters correspond respectively to the slope of the axial stress-axial strain curve and to the opposite of the slope of the radial strain-axial strain curve.

A representative value of unconfined compressive strength for a given rock can be difficult to obtain as it can vary widely within the same rock mass. Considering this large variability, a *UCS* obtained by testing just a few samples cannot be considered as totally reliable.

Various phenomena are involved. Firstly, the friction at the area of contact between the platens generates a non-uniform distribution of the stresses in the sample. To reduce these edge effects, the sample height must be at least twice the diameter. The moisture content may also have a significant effect on the measurement, due to the presence of capillary bridges. Lastly, this type of test is highly sensitive to the presence of the slightest heterogeneity or crack in the sample. Note that it would be more appropriate to measure *UCS* using a triaxial device on a saturated sample subjected to very low confinement. Table 1.2.1 gives the orders of magnitude of unconfined compressive strengths for rocks frequently encountered in petroleum engineering.

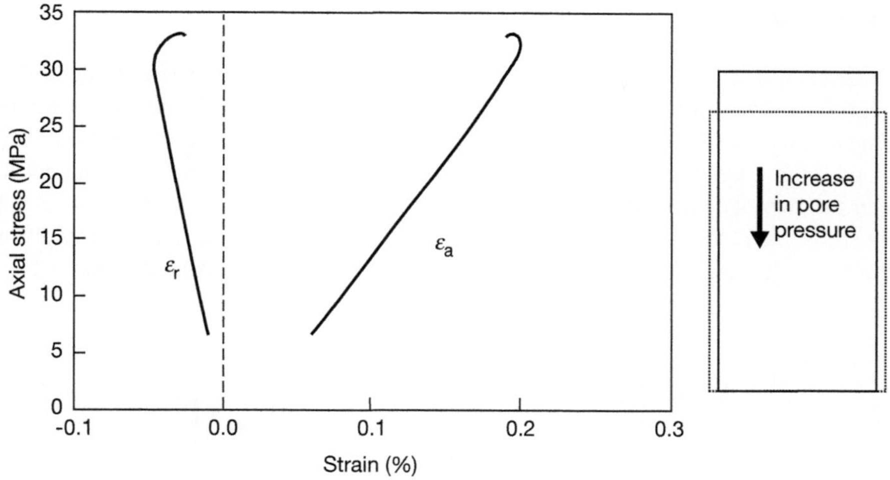

Figure 1.2.1

Unconfined compression test on a sample of Berea sandstone with 24% porosity.

Table 1.2.1 Order of magnitude of the static elastic properties of rocks with different *UCS*s [partially from Le Tirant and Gay, 1972 and Fjaer *et al.*, 1992].

Rock type	*UCS* (MPa)	*E* (GPa)	υ
Poorly consolidated sandstone	< 10	< 1	0.4
Poorly cemented sandstone	10 to 50	5 to 10	0.3 – 0.35
Consolidated sandstone	50 to 250	10 to 30	0.1 to 0.35
Highly consolidated sandstone	> 100	30 to 50	0.15 to 0.25
Hard and very hard limestone	100 to 200	50 to 100	0.2 to 0.3
Soft limestone marl	2.5 to 5	6 to 10	0.4 to 0.45
Soft shale	0.5 to 5	1 to 5	0.3 to 0.4
Highly consolidated shale	5 to 100	5 to 70	0.2 to 0.3
Salt	10 to 30	15 to 40	0.3 to 0.4

1.2.2.2 Uniaxial tension test

Tensile strength describes the ability of the rock to resist tensile stress. The direct measurement of tensile strength can be performed by gluing the opposite ends of a cylindrical rock sample to metal jaws which are fixed to the platens of a test machine. This type of test is rarely carried out, being difficult to implement and interpret. The tensile strength measured in this way is highly dependent on the sample size and the stress concentrations generated by the techniques used to attach the rock to the platens of the press. Consequently, to estimate the tensile strength of a rock the samples must be chosen carefully and a large number of tests conducted.

1.2.2.3 Brazilian test

The Brazilian test is an indirect experimental method to measure the tensile strength of rock. Although it is more frequently used than the uniaxial tension test, being much simpler to implement, it has the disadvantage of introducing non-homogeneous stresses. It is carried out by compressing a cylindrical rock sample of height/diameter ratio less than 1 diametrically between the two platens of a load frame (Figure 1.2.2). The failure occurs under this load along a stress crack, locally roughly in the diametral loading plane. The tensile strengths measured are reproducible and consistent with the values obtained in uniaxial tension.

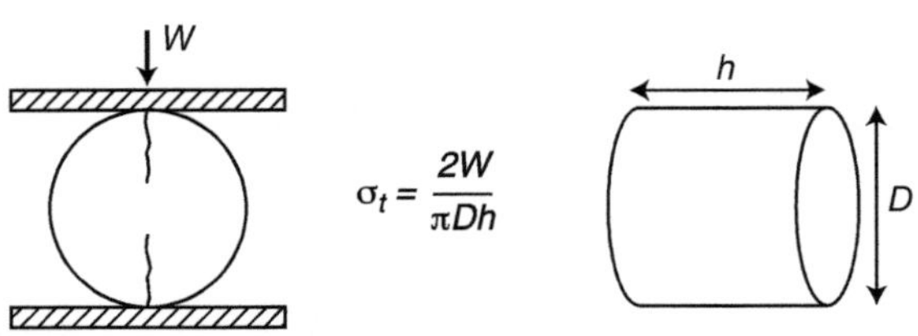

Figure 1.2.2

Principle of the Brazilian test.

In practice, to estimate the tensile strength, the rule is to take a ratio of 8 to 10 with UCS.

1.2.3 Uniaxial strain (or oedometric) test

The oedometric test is used to determine the uniaxial compressibility of a soil- or rock-type material.

The oedometric cell is a "non-deformable" cylindrical mould that can contain a thin sample (typically 20 mm high by 40 mm diameter), placed between two highly porous discs and compressed with a piston (Figure 1.2.3). The horizontal strains being prevented, the sample is in a state of uniaxial strain. The axial strain is then exactly equal to the volumetric strain. The porous discs placed at each end of the sample ensure uniform distribution of the upstream and downstream pore pressures. These pressures can be controlled independently. A differential pressure and therefore a flow can be applied between the upstream and downstream ends of the sample, to evaluate in particular its vertical permeability.

The oedometric test consists in applying an increasing axial load on the sample, usually in stages, using a lever arm loaded with weights. Each stage is maintained until the axial strain of the sample has stabilised. The test can either be drained (upstream and downstream pore pressures controlled) or undrained (sample hydraulically isolated, pore circuits closed).

At regular intervals defined by the operator the following quantities are recorded:

- axial stress (force sensor),
- radial stress (oedometric cells equipped with strain gauges),
- upstream and downstream pore pressures (pressure transducers),

- piston displacement (Linear Variable Displacement Transducer), used to determine the axial strain of the sample,
- temperature (it is recommended to place the oedometers in an air-conditioned room since the pressures imposed are highly sensitive to temperature variations).

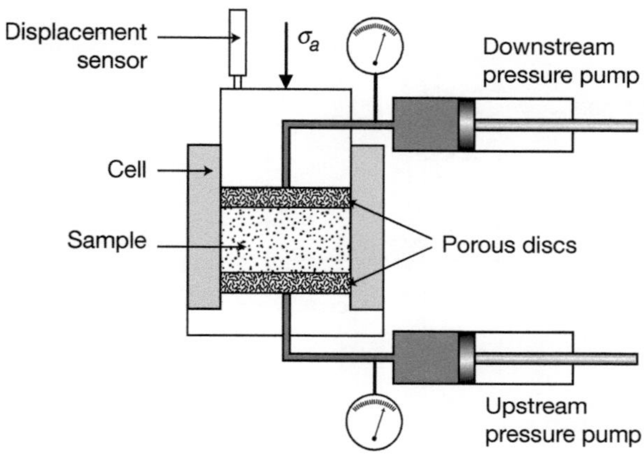

Figure 1.2.3

Diagram of an oedometric cell.

These tests are particularly well-suited to the study of elastoplastic behaviour of very poorly permeable rocks. Since the sample is thinner than for a triaxial test, pressures reach equilibrium more quickly. Note that, using oedometers, it is not generally possible to apply loads large enough to cause sample failure.

Oedometric tests can be used to determine:

- the drained uniaxial compressibility,
- the creep properties (axial strain under constant load as a function of time),
- several poromechanical parameters [Bemer *et al.*, 2001],
- the axial permeability.

1.2.4 Conventional triaxial test

The conventional triaxial test is carried out on cylindrical rock samples subjected to an axial load, a uniform radial stress (via a pressurised confinement fluid) and an interstitial pressure in the interconnected porous network. The sample is surrounded by an impermeable jacket preventing any communication between the pore fluid and the confinement fluid. The pore pressure must always be less than the confining pressure. The imposed stress field is therefore axisymmetrical and defined by the axial stress σ_a, the confining pressure p_c and the pore pressure p_p (Figure 1.2.4). This load is associated with a strain state defined by the axial strain ε_a and the radial strain ε_r.

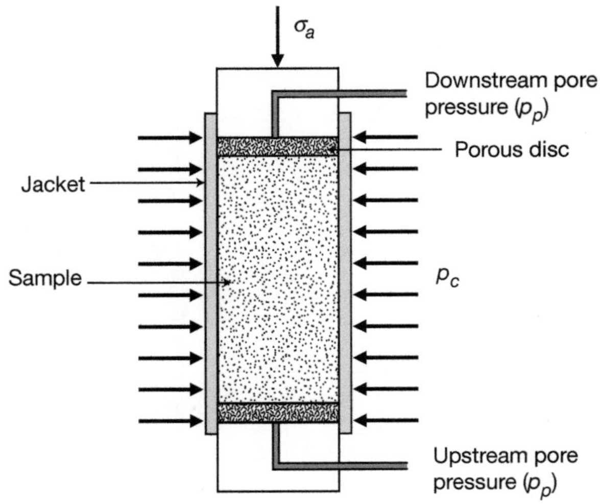

Figure 1.2.4

Stress and pressure states during a triaxial test.

The piston applying the axial stress can be loaded mechanically using a load frame or hydraulically via an upper pressure chamber. The confining and pore pressures are imposed by servo-controlled hydraulic pumps. The axial and radial stresses are controlled independently. Consequently, any required stress path can be applied to the specimen mounted in the cell.

The devices equipped with two pumps to control the pore pressure upstream and downstream from the sample can be used to generate a flow through the sample and therefore to measure its vertical permeability. The porous discs placed at each end of the sample ensure uniform distribution of the upstream and downstream pore pressures. Some laboratories have cells that can be used to measure the horizontal permeability of the sample under triaxial load.

Various devices can be used to measure the axial and radial strains. A local measurement of these two strains can be obtained using gauges stuck to the sample. Vertical displacement sensors (LVDT or cantilever type) placed inside the confinement chamber will provide a semi-local measurement of the axial strain (over much of the sample height). It is not recommended to place these sensors outside the cell, since the global measurement carried out also includes the cell strains, sensitive in particular to the confining pressure applied. The radial strain can also be measured using horizontal displacement sensors (LVDT or cantilever type) placed on the jacket half-way up the sample (local measurement) or a small chain around the sample on a diameter (semi-local measurement). Note that the jacket strains must be calibrated in order to use these two measuring instruments correctly.

The results obtained are often studied in terms of mean stress p and deviatoric stress q, associated with the volume strain ε_v and the deviatoric strain ε_d:

$$p = \frac{\sigma_a + 2p_c}{3} \qquad \varepsilon_v = \varepsilon_a + 2\varepsilon_r \tag{1.2.1}$$

$$q = \sigma_a - p_c \qquad \varepsilon_d = \frac{2}{3}\left(\varepsilon_a - \varepsilon_r\right) \tag{1.2.2}$$

The sample can be isolated by two valves located on the upstream and downstream pore circuits. When these valves are open and the pore pressure is kept constant via a servo-controlled pump, the test is said to be "drained". When they are closed, the volume of fluid inside the sample remains unchanged and the test is said to be "undrained". Note that, in this case, the pore pressure is controlled by the stress state imposed and the rock behaviour.

The first step of a triaxial test consists in saturating the sample with the test fluid, either mineral oil (which offers the advantage of avoiding any chemical interaction with the rock constituents), water, brine, etc. The test generally starts by applying a high vacuum in the sample previously placed in the triaxial cell. Then, under low initial axial stress and confining pressure, an upstream pore pressure (below the confining pressure) is applied, the downstream valve being kept closed. At this point, we simply monitor the volume of fluid injected; the sample is considered to be saturated when the volume reaches a constant value. Note that with a reservoir rock, this procedure assumes that the sample has first been cleaned by removing all the *in situ* fluids. To date, no thorough studies have been conducted to determine the possible impact of this cleaning on the hydromechanical properties of the rock. Note, however, that the rock wettability properties may be modified by the chemical treatment applied.

The standard loading path is a drained path with constant pore pressure. The stresses are increased first isotropically (hydrostatic loading) up to the required confining pressure, then only the axial stress is further increased up to sample failure.

The axial stress is increased at constant axial strain rate (velocity control) or at constant axial load rate $\mathrm{d}\sigma_a / \mathrm{d}t$ (force control). Unloading/reloading cycles can be carried out while loading to estimate the elastic properties.

More complex loading paths can be performed, in particular to simulate the effects of depletion on the behaviour of a reservoir (see § 3.1.3). An uniaxial strain test can be carried out if the radial strain rate $\mathrm{d}\varepsilon_r / \mathrm{d}t$ is maintained at zero via a servo-control system. A variation of the confining pressure proportional to that of the axial pressure can also be applied.

When a constant axial load is imposed, a creep test is carried out; when a constant axial strain is imposed, a relaxation test is carried out.

1.2.5 Effective stresses and drained triaxial test

Remember that the effective stress is given by $\underline{\underline{\sigma}}' = \underline{\underline{\sigma}} - \alpha\, p_p \underline{\underline{1}}$.

By definition, the deviatoric stress is "effective", it does not depend on the pore pressure:

$$q' = \sigma_a' - p_c' = \left(\sigma_a - \alpha\, p_p\right) - \left(p_c - \alpha\, p_p\right) = \sigma_a - p_c = q$$

We will now consider the results of a standard drained triaxial test carried out at constant pore pressure. The end of hydrostatic loading is characterised by a mean stress p_c^o, zero deviatoric stress and a pore pressure p_p^o. This point is generally chosen as reference state when calculating strains. The stress-strain relations then concern the stress increments:

$$q = \sigma_a - p_c^o \text{ and } \Delta p = p - p_c^o = \frac{\sigma_a + 2p_c^o}{3} - \frac{3p_c^o}{3} = \frac{q}{3} \qquad (1.2.3)$$

Since the pore pressure is constant, irrespective of the expression of the effective stress considered, we obtain:

$$\Delta p' = \Delta p$$

In this precise case, therefore, characterisation of the stress-strain relations is independent of the expression of the effective stress considered.

To determine the failure criterion which defines a domain in the stress space outside which the rock cannot withstand the loads, various experimental studies have demonstrated that, provided that the permeability is sufficient to ensure a homogeneous pore pressure distribution, rock failure is governed by Terzaghi's effective stress:

$$\underline{\underline{\sigma}}' = \underline{\underline{\sigma}} - p_p \underline{\underline{1}} \qquad (1.2.4)$$

If we plot the values at failure obtained at various pore pressures on a $p - q$ plane (p = mean stress, q = deviatoric stress), we obtain a failure curve for each pore pressure level (Figure 1.2.5). If we plot the same values in a $p' - q$ plane (p' = Terzaghi's mean effective

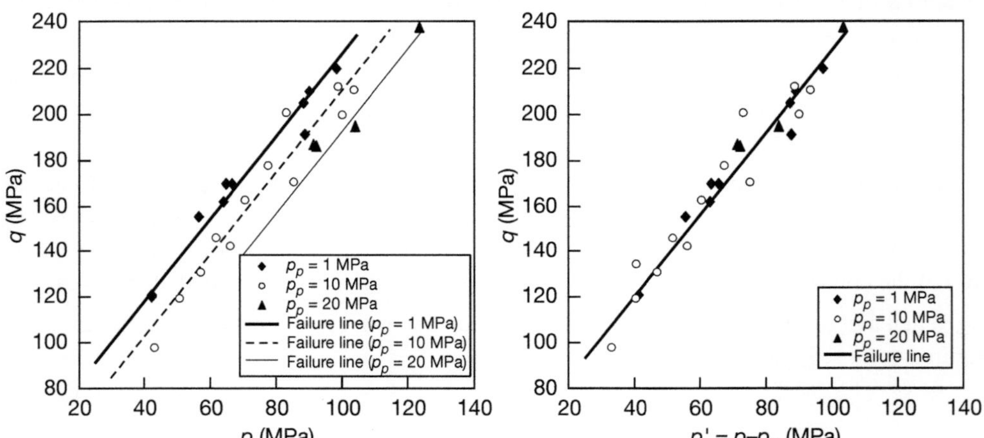

Figure 1.2.5

Deviatoric stress at failure as a function of mean stress and mean effective stress on a limestone with 10% porosity [Vincké *et al.*, 1998].

stress, q = deviatoric stress), the various curves lie on top of each other. In other words, increasing the pore pressure by Δp_p will produce the same reduction in axial stress at failure as reducing the confining pressure by Δp_p [Vincké *et al.*, 1998].

The Mohr-Coulomb failure criterion traditionally used to describe brittle failure of rocks can therefore be written in the form:

$$q = A + B\,p' \text{ where } p' = p - p_p \tag{1.2.5}$$

parameters A and B being related to the (drained) cohesion c' and the (drained) internal angle of friction φ' of the rock by the following relations:

$$A = \frac{6\cos\varphi'}{3 - \sin\varphi'}c' \qquad B = \frac{6\sin\varphi'}{3 - \sin\varphi'}$$

Since we must reason in terms of global stress level applied to the rock in order to study the failure, Terzaghi's effective stress is generally adopted to represent all the results obtained during a triaxial test. For a loading path with pore pressure variations, different effective stresses may need to be considered, including Biot's effective stress which is related to the rock elastic strains:

$$\underline{\underline{\sigma}}' = \underline{\underline{\sigma}} - b\,p_p \underline{\underline{1}} \tag{1.2.6}$$

where b is Biot's coefficient.

1.2.6 Rock behaviour during compression

The behaviour of a rock sample subjected to compression is highly dependent on:

- the rock type (sandstone, carbonate, stiff clays, shale, etc.),
- the initial rock porosity level,
- the loading path followed;

and various microstructural parameters such as, for the sandstones: grain size, cementation, clay content, etc. [Wong *et al.*, 1997]. The saturation state may also have a significant influence, in particular for chalks [Homand and Shao, 2000] [Matà, 2001].

This paragraph is largely based on the experimental studies conducted by Wong, Zhu and David [Wong *et al.*, 1997] [Zhu and Wong, 1997] for sandstones and by Yale and Crawford (1998) for carbonates. Additional data can also be found in [Schutjens and de Ruig, 1996] [Schutjens *et al.*, 1998] [Brawn and Jahns, 1998], etc.

1.2.6.1 Elastic behaviour and measurement of elastic parameters

Elastic behaviour is characterised by perfect reversibility of the strains associated with a load increment. For a large number of rocks, this type of behaviour is encountered for moderate stress ranges.

For a triaxial test, elastic behaviour is observed more especially when loading starts. For small stresses, there may nevertheless be an initial tightening phase corresponding to closure of existing microcracks (generated in particular by sample coring) and associated with a reduction in permeability (Figures 1.2.6 and 1.2.9). Unloading carried out during this phase would result in a slight residual strain. Note that for some rocks, several stress-strain cycles (seasoning) may be required to obtain reproducible behaviour.

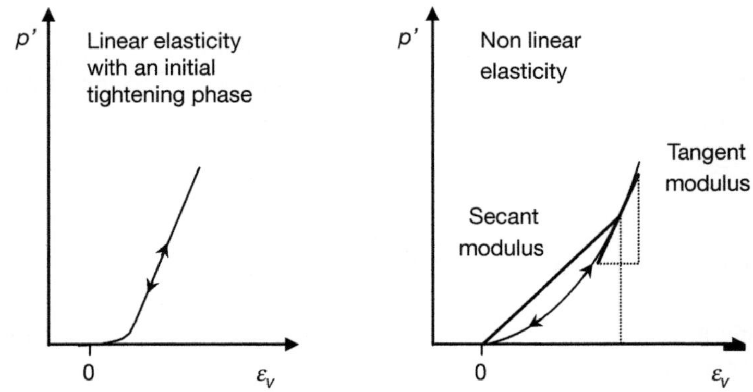

Figure 1.2.6

Stress-strain curves representative of elastic behaviour.

Above a certain level of loading some rocks, including crystalline rocks, sandstones of low to medium porosity, compact limestones, chalks, etc. exhibit linear behaviour. The isotropic elastic parameters of the rock are then easily determined using the stress-strain curves obtained at constant pore pressure (drained test).

The drained bulk modulus K_d is defined by the slope of the mean effective stress-volume strain curve, and the shear modulus G by one third of the slope of the deviatoric stress-deviatoric strain curve:

$$K_d = \mathrm{d}p'/\mathrm{d}\varepsilon_v \qquad (1.2.7)$$

$$3G = \mathrm{d}q/\mathrm{d}\varepsilon_d \qquad (1.2.8)$$

The drained Young's modulus E_d and the drained Poisson's ratio ν_d are deduced from the following relations:

$$E_d = \frac{9K_dG}{3K_d + G} \quad \text{and} \quad \nu_d = \frac{3K_d - 2G}{6K_d + 2G} \qquad (1.2.9)$$

In the special case of hydrostatic loading (i.e. isotropic), only the bulk modulus can be determined.

For purely axial loading, the drained Young's modulus is defined by the slope of the effective axial stress-axial strain curve: $E_d = \mathrm{d}\sigma_a'/\mathrm{d}\varepsilon_v$, and the drained Poisson's ratio by the opposite of the slope of the radial strain-axial strain curve: $\nu_d = -\,\mathrm{d}\varepsilon_r/\mathrm{d}\varepsilon_a$.

Biot's coefficient can be determined by varying the pore pressure, keeping the axial stress and the confining pressure constant. The ratio K_d/b, defined as the opposite of the slope of the pore pressure-volume strain curve, is then measured: $K_d/b = -\mathrm{d}p_p/\mathrm{d}\varepsilon_v$. Knowing K_d, we can deduce b.

Assuming undrained loading, we can calculate the undrained bulk modulus $K_u = \mathrm{d}p/\mathrm{d}\varepsilon_v$, the shear modulus $3G = \mathrm{d}q/\mathrm{d}\varepsilon_d$ and Biot's modulus, provided that Biot's coefficient $Mb = \mathrm{d}p_p/\mathrm{d}\varepsilon_v$ has been determined.

Numerous rocks have nonlinear stress-strain curves in the elastic range, reflecting an increase in bulk modulus as the material is compressed (Figure 1.2.7). This behaviour is observed in particular with rocks whose structure can be schematised by an assembly of cemented "grains". The grains themselves exhibit linear elastic behaviour. The increase in the contact surfaces introduces nonlinearities, however.

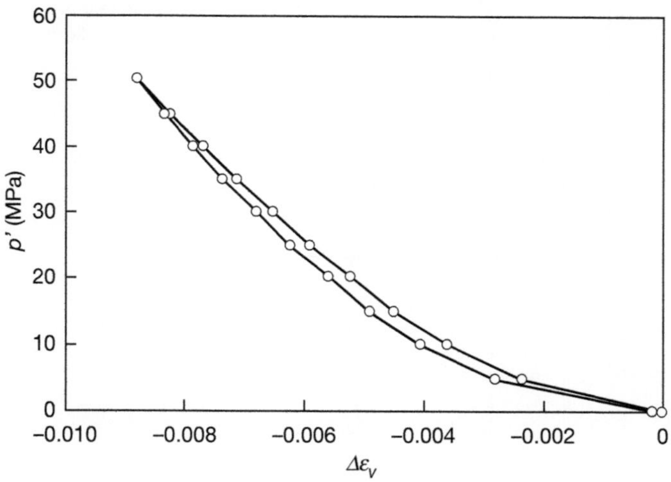

Figure 1.2.7

Drained hydrostatic compression test on a reservoir sandstone with 19.8% porosity.

In practice, nonlinear elasticity curves are interpreted in two ways. Depending on the extent of the stress range associated with the phenomenon studied, the true behaviour may be represented by a linear approximation by determining tangent or secant moduli of elasticity (Figure 1.2.6). The nonlinear nature of the behaviour can also be taken into account via different models: Biot's semilinear model [Bemer *et al.*, 2001] or Hertzian contact model [Vincké, 1994].

1.2.6.2 Sandstones (damage, failure)

Depending on the loading path followed, a sample of sandstone subjected to compression may fail in two different ways.

- At low mean effective stresses, brittle failure through the formation of a shear band crossing the sample (shear localization) is observed. The result is a rapid drop in the ability of the rock to withstand a load.
- At high mean effective stresses, failure occurs under the effect of homogeneous microcracking (cataclastic flow). In this case we speak of ductile failure, with the rock retaining an ability to withstand a load as the strain increases. It is important to note here that this strain is due to slipping along a multitude of intersecting shear planes and is therefore not true plastic strain.

Figure 1.2.8 schematises the corresponding failure surface in the q-p' plane (deviatoric stress-Terzaghi's mean effective stress).

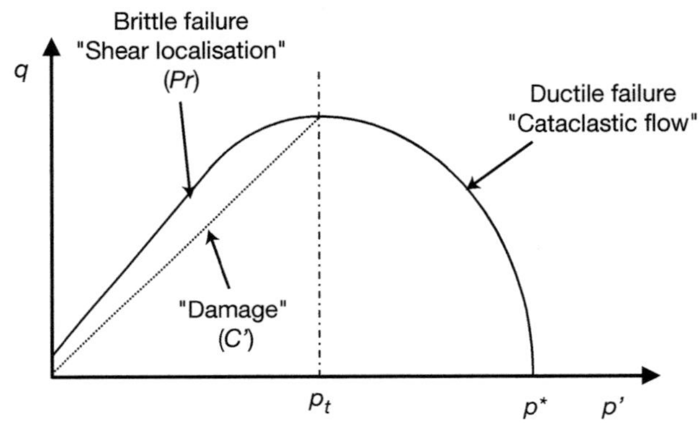

Figure 1.2.8

Failure surface of a sandstone.

Brittle and ductile failure modes are characterised by distinct critical stresses corresponding to different steps in the development of the damage, which is associated with acoustic activity of the rocks: microcracking, slipping with friction and pore collapse, etc. These critical stresses can be identified by measuring the acoustic emissions during the test.

Figure 1.2.9 schematises the change in porosity ($\Delta\Phi$) as the mean effective stress (p') increases depending on the loading path followed. Note that in this case we are considering the variation in Lagrangian porosity, i.e. calculated with respect to the initial volume of the sample. The various curves show a tightening phase followed by an elastic phase (assumed linear on the figure), these two phases being associated with a reduction in the permeability. The third phase is nonlinear and occurs after crossing a critical effective pressure which depends on the loading path (C', C^* or p^*). The phenomena involved in this phase, known as the post-elastic phase, are complex and may lead to a decrease or an increase in the permeability reflecting the competition under the effect of progressive microcracking of the rock between firstly the greater connectivity and secondly the greater tortuosity and/or reduction in the size of the pore throats.

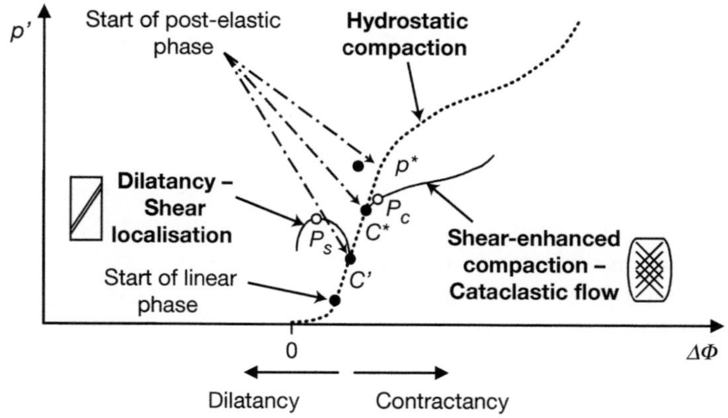

Figure 1.2.9

Behaviour of a sandstone subjected to compression.

Hydrostatic loading

The curve $\Delta\phi - p'$ presents a point of inflection at the critical effective pressure p^*, associated with the onset of grain crushing and pore collapse induced by the growth of transgranular cracks and is accompanied by an increase in acoustic activity indicating a change of physical mechanism. As long as p^* is not reached, the rock behaviour corresponds to elastic reduction of the porosity. Above p^*, the post-elastic behaviour phase starts, systematically inducing a decrease in the permeability in case of hydrostatic compression. The notion of low or high mean stresses is defined with respect to the position of the ratio p'/p^* on the scale [0,1].

Ductile failure (failure by homogeneous microcracking or cataclastic flow)

Ductile failure is observed for deviatoric loads at high mean effective stress. Compared with hydrostatic loading, the curve $\Delta\Phi - p'$ indicates a higher reduction in the porosity above an effective compaction pressure C^*, below p^*. This higher reduction in the porosity, together with a reduction in permeability more pronounced than in the hydrostatic case, is induced by the deviatoric part of the stress field. We then speak of shear-enhanced compaction. The curve $\Delta\Phi - p'$ presents a point of inflection at the effective collapse pressure P_c.

As long as C^* is not reached, the rock behaviour corresponds to an elastic reduction of the porosity independent of the deviatoric stress. Above C^*, the acoustic emission rate increases more rapidly, indicating the development of intragranular cracks initiated at the points of contact between grains. When they become transgranular, these cracks increase the stresses on the load-bearing grains, finally crushing the grains and causing the pores to collapse. Collapse of the sample occurs at pressure P_c and is associated with a peak in the acoustic emission rate. The non-hydrostatic nature of the loading leads to a preferential orientation of the cracks approximately parallel to the axial stress.

With consolidated sandstones, the stress levels associated with C^* and P_c are very close and only the effective compaction pressure C^* is used to characterise ductile failure.

Brittle failure (failure by creation of a shear band or shear localisation)

Brittle failure is observed for deviatoric loads at low mean effective stress. Compared with hydrostatic loading, the curve $\Delta\Phi - p'$ indicates dilating behaviour above an effective dilatancy pressure C' marking the start of damage and well below p^*. The stress curve then shows a peak P_S, which corresponds to the formation of a shear band crossing the sample, and beyond which we observe "softening" behaviour, indicating the fact that less force is required to strain the sample at constant strain rate. This behaviour phase is sometimes represented by elastoplastic strain softening models. It is important to bear in mind, however, that it does not correspond to a homogeneous strain in the rock, but to slipping of rigid blocks at the fractures.

As long as C' is not reached, the rock behaviour corresponds to an elastic reduction of the porosity independent of the deviatoric stress. Intergranular cracks develop between C' and P_S (probably due to cementation failure), inducing a faster increase in the acoustic emission rate. These cracks, whose preferential orientation is approximately parallel to the axial stress, allow relative movement of the grains thereby causing an overall dilatation of the porous space. Performing several unloading/reloading cycles between C' and P_S would demonstrate a reduction in the rock stiffness due to the effect of the damage. The change in permeability is then related to the initial rock permeability.

- Rocks of low initial permeability, i.e. poorly connected rocks, will be highly sensitive to the increased connectivity, due to the development of microcracks, and their permeability will therefore increase.
- For rocks of high initial permeability, i.e. well-connected rocks, the microcracks do not modify the overall connectivity but allow grain movements, which increase the tortuosity and/or reduce the size of the pore throats, thereby reducing the permeability.

Intragranular cracks develop on approaching the stress peak; subsequent coalescence of these cracks finally results in failure through formation of a shear band crossing the sample. The bulk permeability is then controlled by the properties of the localisation zone, which may in particular obstruct the flow of fluids. The stress peak is marked by a peak in the acoustic emission rate. The maximum acoustic emission rate associated with a brittle failure (dilating strain) is less than the maximum rate associated with a ductile failure (compacting strain).

1.2.6.3 Carbonates (plasticity, failure)

Unlike sandstones, carbonates may develop a significant "plastic" behaviour before failure. The term plastic refers here to a behaviour showing irreversible strains without a decrease of the elastic properties. These strains are induced by the coupling of the microfracturing (associated with crushing and frictional slipping) and the plasticity of the crystals. Plasticity is essentially encountered for loading paths at high effective mean stress and for rocks of high initial porosity, in particular chalks.

In addition, since very few events generate acoustic activity in carbonates, the effective critical pressures must be determined using the stress-strain curves only, creating a larger dispersion than for the sandstones [Vajdova *et al.*, 2004].

This paragraph is based on the experimental results obtained by Yale and Crawford (1998) on various oil-saturated carbonates of porosities ranging from 14% to 42%. The tests followed various stress paths to approximate the *in situ* conditions corresponding to extraction of oil from a reservoir.

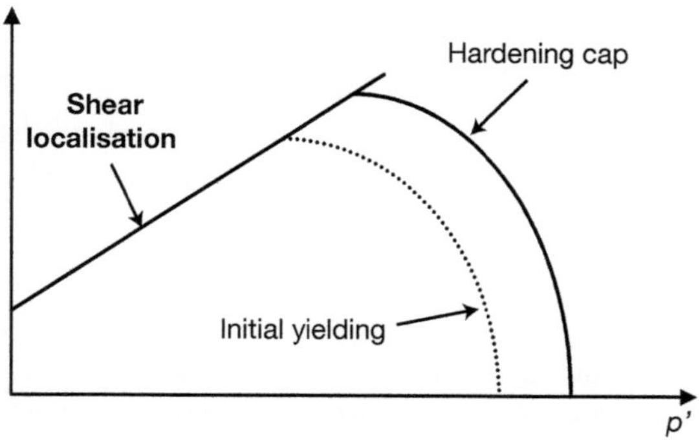

Figure 1.2.10

Carbonate failure/plastification model.

The stress paths with low mean effective stress resulted in brittle failure of the samples. The stress paths with higher mean effective stress were used to determine two yield surfaces (Figure 1.2.10). Engstrøm (1991) also observed the existence of progressive strain-hardening of chalks for this type of loading.

According to Yale and Crawford (1998), the first yield surface (initial yielding) corresponds to the stress state from which the curve $\Delta\Phi - p'$ deviates from its initial linear slope by more than 10%. This surface is to be compared with the effective compaction pressures C^* associated with sandstones.

The second yield surface (hardening cap) is reached at the point of inflection of the curve $\Delta\Phi - p'$ and is therefore to be compared with the collapse pressures P_c. The fundamental difference between sandstones and carbonates lies in the significant gap between the two yield surfaces for carbonates: the ratio between the critical hydrostatic pressures of hardening cap and initial yielding reaches 1.8 for 20% porosity carbonates.

Strain hardening of the samples is positive (irreversible strain can only increase by increasing the load) and mainly controlled by the porosity of the carbonates tested: for low-porosity samples, the initial hardening threshold stresses are higher and the gap between the initial yielding and the hardening is larger. In addition, dispersion of experimental results is more pronounced for higher-porosity samples due to the greater sensitivity to small-scale heterogeneities.

In case of brittle failure, the stress levels associated with damage onset and the failure peak are very close for carbonates. Failure therefore occurs quite suddenly after the elastic phase. As for sandstones and with the same reserves, the post-peak behaviour is sometimes represented by elastoplastic strain-softening models.

1.2.6.4 Shales

The term shale designates rocks whose clay mineral content (clay volume divided by total volume) is greater than 35% [Vernik *et al.*, 1993] [Plumb, 1994]. It therefore includes rocks of highly variable composition and structure. Clay minerals consist of stacks of elementary sheets. Depending on the type of clay, the bonds between sheets are more or less strong. Unlike kaolinite and illite, smectite has weak bonds, making it highly sensitive to water content; as a result it has the ability to fix water molecules between two neighbouring sheets and exhibits a high swelling potential.

The study of swelling clays is highly complex and will not discussed here.

Very broadly speaking, we can identify two main categories of shales according to their water content (weight of water divided by weight of solid particles) [Rousset, 1988]:

- "plastic" shales with high water content (> 15 %),
- "stiff" shales with low water content (< 10%).

Since the elastic limit of plastic shales is virtually zero, they can withstand high strains with no apparent macroscopic failure. After a strain hardening phase, loadings at low mean effective stress lead to a "rounded" stress peak, followed by softening behaviour. We then obtain a stress plateau demonstrating the residual strength retained by the shale.

This type of behaviour is generally represented by a Cam-Clay type yield surface elastoplastic model (Figure 1.2.11).

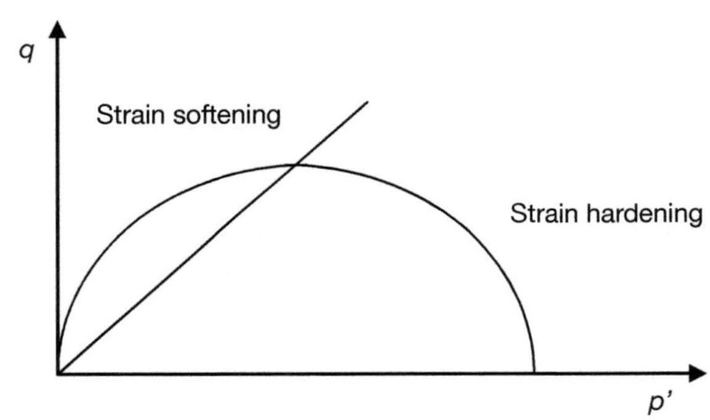

Figure 1.2.11

Cam-Clay model yield surface.

The behaviour of stiff shales is more like that of an elastobrittle rock. They may even exhibit very high strengths [Lin, 1981]. At low mean stresses, a sharp peak is observed, followed by a rapid drop in the stress. The brittle failure data of stiff shales can be represented by a linear Mohr-Coulomb criterion or a nonlinear criterion [Green *et al.*, 1973] [Lin, 1981] [Sargand and Hazen, 1987] [Cook *et al.*, 1991] [Steiger and Leung, 1992] [Gavito, 1996]. Few experimental data are available, however, to close the yield surface for high mean stresses.

In addition, plastic shales exhibit pronounced delayed response. At a given stress state, the rock may therefore undergo strains over a long period of time, which may stabilise or not. Stiff shales may also develop more or less large delayed strains. Viscous behaviour models may be required to take into account these phenomena [Boidy, 2002].

1.2.7 Simulating depletion in the laboratory

Depletion tests in the laboratory are generally conducted on cylindrical samples placed in a triaxial cell, with test procedures leading to a stress state more or less close to the *in situ* conditions. Depletion can be simulated by either reducing the pore pressure at constant axial stress to simulate a constant weight of earth ("true depletion" tests), or by increasing the effective stresses at constant pore pressure. Obviously, simulating depletion by reducing pore pressure is more representative of reservoir production mode (see § 3.1.3).

One disadvantage with replacing the pressure reduction by an increase in effective stresses is that it introduces Biot's coefficient. Depletion tests are in fact mainly carried out to characterise the strain behaviour of the rock and, in this case, the relevant effective stress is Biot's.

The loading path followed in terms of the evolution in axial stress and confining pressure may be:

- isotropic ($\Delta p_c' = \Delta\sigma_a'$);
- oedometric (i.e. uniaxial, the confining pressure evolution is then controlled automatically to maintain the condition $\Delta\varepsilon_r = 0$);
- proportional ($\Delta p_c' = K\,\Delta\sigma_a'$), in order to follow the reservoir stress path when known (see § 3.1.1).

Numerous experimental studies demonstrate that the behaviour of reservoir rocks is highly dependent on the stress path [Holt, 1990] [Warpinski and Teufel, 1990] [Rhett and Teufel, 1992a] [Ruisten et al., 1996] [Schutjens and de Ruig, 1996] [Keaney et al., 1998] [Schutjens et al., 1998]. Consequently, if the stress path is known, it should be followed in the tests conducted to determine the reservoir rock characteristics.

1.3 MODELLING ROCK BEHAVIOUR

Warning! Throughout Chapter 1.3, compressive stresses are taken to be negative, according to the solid mechanics convention.

We will examine the modelling of various aspects of the behaviour of a rock saturated with a single fluid: a liquid or a perfect gas. Firstly, we will consider the case of a solid skeleton exhibiting isotropic linear elastic behaviour. We will then introduce nonlinear elastic behaviour of the solid skeleton. Our intention here is not to provide an exhaustive description of the theoretical framework of poromechanical modelling, which interested readers can find in [Coussy, 1991, 1995, 2004], but to show precisely where the nonlinearities occur in the various cases considered. Lastly, we will present an example of a poroelastoplastic behaviour model: the modified Cam-Clay model.

1.3.1 Representation of the porous medium

We adopt the formalism developed by Coussy (1991, 1994, 2004). The porous medium is represented as the superimposition of two continuous interacting media: the solid skeleton and the interstitial fluid (Figure 1.3.1). It is considered as an open continuous medium and its strains are described by those of its skeleton.

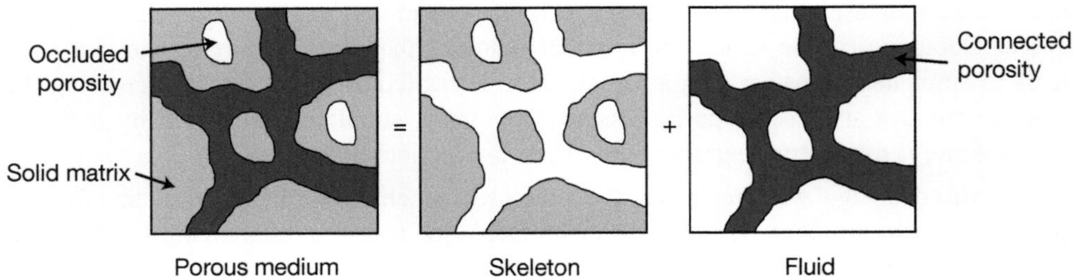

Figure 1.3.1

Porous medium seen as the superimposition of two continuous media.

1.3.2 Fundamental assumptions

We make the following assumptions:

- quasistatic evolutions (to neglect inertia forces);
- infinitesimal transformations ($\left\|\underline{grad}\,\underline{\xi}\right\| << 1$, where $\underline{\xi}$ is the displacement vector of the skeleton particles, to ensure first order agreement of the Lagrangian and Eulerian approaches);
- small displacements of the skeleton particles with respect to a characteristic length of the porous medium L_o ($\left\|\underline{\xi}/L_o\right\| << 1$, to merge the initial and current configurations for the space arguments of the various fields);

and possibly:

- small variations in the fluid density ($\left|\left(\rho_{fl} - \rho_{fl}^o\right)/\rho_{fl}^o\right| << 1$);
- small variations in the Lagrangian porosity ($\left|\left(\Phi - \Phi_o\right)/\Phi_o\right| << 1$).

1.3.3 Equilibrium equations

1.3.3.1 Conservation of mass

The continuity equation expresses conservation of the fluid mass:

$$\frac{dm}{dt} + div\,\underline{w} = 0 \tag{1.3.1}$$

The fluid mass content per initial volume unit m is simply related to the Lagrangian porosity Φ by the following relation which expresses complete saturation of the medium:

$$m = \rho_{fl}\,\Phi \tag{1.3.2}$$

where ρ_{fl} designates the fluid density.

The relative flow vector of fluid mass $\underline{w}$ is directly proportional to the relative flow rate of the fluid with respect to the skeleton:

$$\underline{w} = \rho_{fl}\,\underline{V} \qquad \underline{V} = \phi\left(\underline{v_{fl}} - \underline{v_s}\right) \tag{1.3.3}$$

where ϕ is the Eulerian porosity, which is the ratio of the current porous volume to the current total volume, and $\underline{V}$ is the filtration vector.

Assuming infinitesimal transformations, the Lagrangian and Eulerian porosities are simply related by:

$$\Phi = (1+\varepsilon)\phi$$

where $\varepsilon = \mathrm{tr}\,\underline{\underline{\varepsilon}}$.

1.3.3.2 Equation of motion

Momentum balance leads to the equation of mechanical equilibrium:

$$\underline{div\,\underline{\underline{\sigma}}} + \rho\,\underline{F} = \underline{0} \tag{1.3.4}$$

where $\rho = \rho_s(1-\phi) + \rho_{fl}\,\phi$ is the density of the porous medium and $\underline{F}$ represents the external volume forces often reduced to the gravitational forces $\underline{g}$.

Assuming small variations in fluid density and porosity, the equation of mechanical equilibrium can be simplified to:

$$\underline{div\,\underline{\underline{\sigma}}} + \rho_o\,\underline{F} = \underline{0}$$

where $\rho_o = \rho_s^o(1-\Phi_o) + \rho_{fl}^o\Phi_o$.

1.3.4 Fluid constitutive equations

The fluid constitutive equations can be defined by applying the first and second laws of thermodynamics to the interstitial fluid. In particular, they can be expressed using the fluid-specific free enthalpy g_{fl}. In this case, they relate the density ρ_{fl} and the fluid-specific entropy s_{fl} to the thermodynamic pressure p_p (associated with the pore pressure) and to the temperature T.

$$dg_{fl} = \frac{dp_p}{\rho_{fl}} - s_{fl}\,dT \tag{1.3.5a}$$

or

$$g_{fl} = g_{fl}(p_p, T)$$

$$\frac{1}{\rho_{fl}} = \frac{\partial g_{fl}}{\partial p_p} \qquad -s_{fl} = \frac{\partial g_{fl}}{\partial T} \tag{1.3.5b}$$

Note that the fluid-specific enthalpy is related to its specific free energy ψ_{fl}:

$$g_{fl} = \psi_{fl} + \frac{p_p}{\rho^{fl}} \tag{1.3.6}$$

By differentiating the constitutive equations (1.3.5b), we obtain:

$$\frac{d\rho_{fl}}{\rho_{fl}} = \frac{dp_p}{K_{fl}} - 3\alpha_{fl}\, dT \tag{1.3.7a}$$

$$ds_{fl} = -3\alpha_{fl} \frac{dp_p}{\rho_{fl}} + c_p \frac{dT}{T} \tag{1.3.7b}$$

where K_{fl} designates the fluid bulk modulus, $3\alpha_{fl}$ the fluid volumetric thermal expansion coefficient and c_p the fluid volumetric specific heat capacity at constant pressure.

For a low compressibility liquid and small pressure and temperature variations, we can obtain a linearised expression of the fluid constitutive equations by integrating equations (1.3.7) with K_{fl}, $3\alpha_{fl}$ and c_p constant:

$$\frac{1}{\rho_{fl}} = \frac{1}{\rho_{fl}^o}\left[1 - \frac{p_p - p_p^o}{K_{fl}} + 3\alpha_{fl}\left(T - T_o\right)\right]$$

$$s_{fl} = s_{fl}^o - 3\alpha_{fl} \frac{p_p - p_p^o}{\rho_{fl}^o} + c_p \frac{T - T_o}{T_o}$$

where the index o refers to the initial state.

This linearisation is no longer relevant, however, for a high compressibility gas. If this gas is represented by a perfect gas model, we obtain:

$$\rho_{fl} = \rho_{fl}^o \frac{p_p}{p_p^o} \frac{T_o}{T} \tag{1.3.8}$$

Differentiation of equation (1.3.8) and comparison with equation (1.3.7a) demonstrate that the bulk modulus and the volumetric thermal expansion coefficient of a perfect gas are equal respectively to: $K_{fl} = p_p$ and $3\alpha_{fl} = 1/T$.

1.3.5 Clausius-Duhem inequality and thermal equation

By applying the first law of thermodynamics to the porous medium seen as the superimposition of the solid skeleton and the interacting interstitial fluid, we obtain the energy equation, which expresses the conservation of energy:

$$\frac{dE}{dt} = \underline{\underline{\sigma}} : \frac{d\underline{\underline{\varepsilon}}}{dt} - div\left(h_{fl}\,\underline{w} + \underline{q}\right) + \underline{F}.\underline{w} \tag{1.3.9}$$

where E designates the internal energy density of the porous medium per initial volume unit, $h_{fl} = g_{fl} - T\,s_{fl}$ the specific enthalpy of the fluid and $\underline{q}$ the outgoing heat flow vector.

The second law of thermodynamics leads to the entropy balance:

$$\frac{dS}{dt} + div\left(s_{fl}\,\underline{w}\right) \geq - div\left(\frac{\underline{q}}{T}\right) \tag{1.3.10}$$

where S designates the entropy density of the porous medium per initial volume unit. The ≥ sign shows that there is a spontaneous production of entropy: φ/T, where φ is the overall dissipation.

By introducing the porous medium density of free energy $\psi = E - T\,S$, using the above two relations, we derive the Clausius-Duhem inequality which shows that the dissipation φ is non-negative:

$$\varphi = \varphi_s + \varphi_{fl} + \varphi_{th} \geq 0 \tag{1.3.11}$$

$$\varphi_s = \underline{\underline{\sigma}} : \frac{d\underline{\underline{\varepsilon}}}{dt} + g_{fl}\frac{dm}{dt} - S\frac{dT}{dt} - \frac{d\psi}{dt} \tag{1.3.12}$$

represents the intrinsic dissipation, which turns out to be related to the solid skeleton alone. If we introduce the entropy and free energy of the skeleton per initial volume unit, $S_s = S - m\,s_{fl}$ and $\psi_s = \psi - m\,\psi_{fl}$, the fluid constitutive equations (1.3.5a) and the relation $m = \rho_{fl}\,\Phi$ lead to the following expression of the intrinsic dissipation:

$$\varphi_s = \underline{\underline{\sigma}} : \frac{d\underline{\underline{\varepsilon}}}{dt} + p_p\frac{d\Phi}{dt} - S_s\frac{dT}{dt} - \frac{d\psi_s}{dt} \tag{1.3.13}$$

$$\varphi_{fl} = \left[- \underline{grad}\,p_p + \rho_{fl}\underline{F}\right]\cdot\underline{V} \tag{1.3.14}$$

represents the dissipation associated with the fluid and more particularly its movement with respect to the skeleton.

$$\varphi_{th} = -\frac{\underline{q}}{T}\cdot\underline{gradT} \tag{1.3.15}$$

represents the thermal dissipation.

Due to the highly different nature of these dissipations, we adopt the decoupling assumption which consists in replacing inequality (1.3.11) by three separate inequalities:

$$\varphi_s = \underline{\underline{\sigma}} : \frac{d\underline{\underline{\varepsilon}}}{dt} + p_p \frac{d\Phi}{dt} - S_s \frac{dT}{dt} - \frac{d\psi_s}{dt} \geq 0 \tag{1.3.16}$$

$$\varphi_{fl} = \left[-\underline{grad}\, p_p + \rho_{fl} \underline{F} \right] \cdot \underline{V} \geq 0 \tag{1.3.17}$$

$$\varphi_{th} = -\frac{\underline{q}}{T} \cdot \underline{grad}\, T \geq 0 \tag{1.3.18}$$

After identifying the spontaneous production of entropy φ/T, inequality (1.3.10) finally leads to the thermal equilibrium equation:

$$T\left[\frac{dS}{dt} + div\left(s_{fl}\, \underline{w} \right) \right] = -\, div\, \underline{q} + \varphi_M$$

where $\varphi_M = \varphi_s + \varphi_{fl}$ represents the mechanical dissipation density.

1.3.6 Conduction laws

1.3.6.1 Fluid conduction law

Flow dissipation expression (1.3.17) demonstrates that the law of fluid conduction must relate the filtration vector to the flow force $-\underline{\text{grad}}\, p_p + \rho_{fl} \underline{F}$. In this case, its simplest form is the linear relation given by Darcy's law. For an isotropic medium, it is written:

$$\frac{\underline{w}}{\rho_{fl}} = \underline{V} = \Lambda \left(-\underline{grad}\, p_p + \rho_{fl} \underline{F} \right) \tag{1.3.19}$$

Note that the volume forces often prove to be negligible compared with the hydraulic gradient.

For a low compressibility interstitial fluid, the hydraulic conductivity Λ is expressed as a function of the intrinsic permeability k, which depends solely on the geometry of the porous network, and of the dynamic viscosity of the interstitial fluid μ_{fl}:

$$\Lambda = \frac{k}{\mu_{fl}} \tag{1.3.20}$$

An additional interstitial pressure dependence must nevertheless be taken into account when the interstitial fluid is a gas [Patron, 1999]:

$$\Lambda = \frac{k}{\mu_{fl}} + \frac{k_g}{p_p} \tag{1.3.21}$$

When these models are included in finite elements simulation software, we may introduce variation laws of the hydraulic parameters k and k_g as a function of the porosity, which represents a simplified way of taking into account the evolution of the porous network and thereby of including this aspect of the hydro-mechanical coupling in the numerical model.

1.3.6.2 Heat conduction law

Thermal dissipation expression (1.3.18) shows that the heat conduction law must relate the heat flow vector to the force responsible for heat production $-\underline{grad}\,T$. In this case, its simplest form is the linear relation given by Fourier's law. For an isotropic medium, it is written:

$$\underline{q} = -\lambda_T\, \underline{grad}\, T \tag{1.3.22}$$

where λ_T is the thermal conductivity.

1.3.7 Constitutive equations of the skeleton and the porous medium

Intrinsic dissipation expression (1.3.13) shows that, due to its nature, the skeleton density of free energy depends on the external state variables: $\underline{\underline{\varepsilon}}$, Φ and T (observable variables whose variation can be controlled externally), their associated variables being $\underline{\underline{\sigma}}$, p_p and S_s. According to the postulate of local state [Coussy, 2004], the skeleton density of free energy is independent of the evolution rates and can finally be expressed in the form:

$$\psi_s = \psi_s\left(\underline{\underline{\varepsilon}}, \Phi, T, \chi_1, \chi_2, \ldots, \chi_N\right) \tag{1.3.23}$$

where the variables χ_i are "internal" state variables (their variation cannot be controlled externally).

Since the intrinsic dissipation cannot be negative, we have:

$$\left(\underline{\underline{\sigma}} - \frac{\partial \psi_s}{\partial \underline{\underline{\varepsilon}}}\right) : d\underline{\underline{\varepsilon}} + \left(p_p - \frac{\partial \psi_s}{\partial \Phi}\right) d\Phi - \left(S + \frac{\partial \psi_s}{\partial T}\right) dT - \frac{\partial \psi_s}{\partial \underline{\chi}} \cdot d\underline{\chi} \geq 0 \tag{1.3.24}$$

We will now consider evolutions where the internal variables are constant. Since inequality (1.3.24) must be satisfied for any evolution in the external variables independently of each other, we obtain the following constitutive equations:

$$\underline{\underline{\sigma}} = \frac{\partial \psi_s}{\partial \underline{\underline{\varepsilon}}} \qquad p_p = \frac{\partial \psi_s}{\partial \Phi} \qquad S = -\frac{\partial \psi_s}{\partial T} \tag{1.3.25}$$

If we now introduce the potential $G_s = \psi_s - p_p \Phi$, the constitutive equations can be written:

$$\underline{\underline{\sigma}} = \frac{\partial G_s}{\partial \underline{\underline{\varepsilon}}} \qquad \Phi = -\frac{\partial G_s}{\partial p_p} \qquad S_s = -\frac{\partial G_s}{\partial T} \tag{1.3.26}$$

Note that equations (1.3.25) and (1.3.26) remain valid for evolutions including a variation of internal variables. Complementary evolution laws governing the evolution of internal state variables need to be introduced however.

Intrinsic dissipation expression (1.3.12) leads to the expression of the density of free energy and the constitutive equations of the porous medium relating the external state variables $\underline{\underline{\varepsilon}}$, m and T to their associated variables $\underline{\underline{\sigma}}$, g_{fl} and S_s:

$$\psi = \psi\left(\underline{\underline{\varepsilon}}, m, T, \chi_1, \chi_2, \ldots, \chi_N\right) \tag{1.3.27}$$

$$\underline{\underline{\sigma}} = \frac{\partial \psi}{\partial \underline{\underline{\varepsilon}}} \qquad g_{fl} = \frac{\partial \psi}{\partial m} \qquad S = -\frac{\partial \psi}{\partial T} \tag{1.3.28a}$$

For porous media with no chemical activity, however, it is simpler to introduce the fluid mass content directly into the constitutive equations (1.3.26):

$$\underline{\underline{\sigma}} = \frac{\partial G_s}{\partial \underline{\underline{\varepsilon}}} \qquad \frac{m}{\rho_{fl}} = -\frac{\partial G_s}{\partial p_p} \qquad S = -\frac{\partial G_s}{\partial T} + m\, s_{fl} \tag{1.3.28b}$$

1.3.8 Linear poroelastic behaviour

1.3.8.1 Skeleton constitutive equations

We will now consider a porous medium whose skeleton exhibits isotropic linear elastic behaviour. We can therefore adopt for ψ_s a quadratic expression ψ_s^{el} which depends solely on the first and second invariants of the strain tensor.

By differentiating the constitutive equations (1.3.26) with $G_s = G_s^{el} = \psi_s^{el} - p_p \Phi$, we obtain the porous skeleton thermoporoelastic constitutive equations which involve only coefficients which are independent of the interstitial fluid:

$$d\underline{\underline{\sigma}} = \left(K_d - \frac{2G}{3}\right) d\varepsilon \, \underline{\underline{1}} + 2G\, d\underline{\underline{\varepsilon}} - b\, dp_p \underline{\underline{1}} - 3\alpha_d K_d\, dT\, \underline{\underline{1}}$$

$$d\Phi = b\, d\varepsilon + \frac{dp_p}{N} - 3\alpha_\Phi dT \tag{1.3.29}$$

$$dS_s = 3\alpha_d K_d\, d\varepsilon - 3\alpha_\Phi\, dp_p + C_\varepsilon \frac{dT}{T}$$

where K_d is the skeleton bulk modulus, G the skeleton shear modulus, b Biot's coefficient, $1/N$ the pore compressibility, $3\alpha_d$ the skeleton volumetric thermal expansion coefficient, $3\alpha_\Phi$ the pore volumetric thermal expansion coefficient and C_ε the skeleton volumetric heat capacity at constant strain.

Assuming that the solid matrix is homogeneous and isotropic, we can determine expressions relating coefficients b, N and α_Φ, to the characteristics of the solid matrix [Coussy, 2004]:

$$b = 1 - \frac{K_d}{K_s} \qquad \frac{1}{N} = \frac{b - \Phi_o}{K_s} \tag{1.3.30}$$

$$\alpha_d = \alpha_s \qquad \alpha_\Phi = \left(b - \Phi_o\right)\alpha_s \tag{1.3.31}$$

where K_s is the bulk modulus of the solid matrix and α_s the thermal expansion coefficient of the solid constituent.

1.3.8.2 Porous medium constitutive equations

The thermoporoelastic constitutive equations of the porous medium considered as an open thermodynamic system involve the fluid mass content m. They are obtained from constitutive equations (1.3.28b) and the fluid constitutive equations (1.3.7).

Differentiating relation (1.3.2), we obtain:

$$\frac{dm}{\rho_{fl}} = d\Phi + \Phi \frac{d\rho^{fl}}{\rho^{fl}}$$

The fluid constitutive equations then lead to the relations:

$$\frac{dm}{\rho_{fl}} = d\Phi + \Phi \frac{dp_p}{K_{fl}} - 3\Phi\alpha_{fl}\, dT$$

$$d\left(m\, s_{fl}\right) = s_{fl}\, dm - 3\Phi\alpha_{fl}\, dp_p + m\, c_p \frac{dT}{T}$$

By introducing the above relations in equations (1.3.29), we can eliminate the Lagrangian porosity variation, replacing it by the variation in fluid mass content. Finally, we obtain the constitutive equations:

$$\begin{aligned}
d\underline{\underline{\sigma}} &= \left(K_d - \frac{2G}{3}\right) d\varepsilon\, \underline{\underline{1}} + 2G\, d\underline{\underline{\varepsilon}} - b\, dp_p \underline{\underline{1}} - 3\alpha_d K_d\, dT\, \underline{\underline{1}} \\
\frac{dm}{\rho^{fl}} &= b\, d\varepsilon + \frac{dp_p}{M} - 3\alpha_m\, dT \\
dS - s_{fl}\, dm &= 3\alpha_d K_d\, d\varepsilon - 3\alpha_m\, dp_p + C_\varepsilon^d \frac{dT}{T}
\end{aligned} \tag{1.3.32}$$

with
$$\frac{1}{M} = \frac{1}{N} + \frac{\Phi}{K_{fl}} \tag{1.3.33}$$

$$\alpha_m = \alpha_\Phi + \Phi \alpha_{fl} \text{ and } C_\varepsilon^d = C_\varepsilon + m c_p \tag{1.3.34}$$

The skeleton characteristics K_d, G, b and α_d, prove to be drained properties of the porous medium, i.e. they can be measured from tests where the interstitial pressure is maintained constant ($dp_p = 0$). The drained volumetric heat capacity at constant strain C_ε^d takes into account the fluid mass content via the term $m c_p$.

Biot's modulus M and the differential thermal expansion coefficient α_m depend on the interstitial fluid. For a porous medium saturated by a low compressibility fluid, Biot's modulus and the differential thermal expansion coefficient can be considered as constant to the first order of approximation. When the interstitial fluid is highly compressible, however, they depend on the interstitial pressure and the temperature.

The drained constitutive equations (1.3.32) can be combined to obtain an undrained expression of these equations:

$$\begin{aligned}
d\underline{\underline{\sigma}} &= \left(K_u - \frac{2G}{3}\right) d\varepsilon \underline{\underline{1}} + 2G d\underline{\underline{\varepsilon}} - bM \frac{dm}{\rho_{fl}} \underline{\underline{1}} - 3\alpha_m K_u dT \\
\frac{dp_p}{M} &= -\, b d\varepsilon + \frac{dm}{\rho^{fl}} + 3\alpha_m dT \\
dS - s_{fl}\, dm &= 3\alpha K_u d\varepsilon - 3\alpha_m M \frac{dm}{\rho_{fl}} + C_\varepsilon^u \frac{dT}{T}
\end{aligned} \tag{1.3.35}$$

with $K_u = K_d + M b^2$, $3\alpha_u K_u = 3\alpha_d K_d + 3\alpha_m M b$ and $C_\varepsilon^u = C_\varepsilon^d - 9T\alpha_m^2 M$. K_u is the undrained bulk modulus and $3\alpha_u$ the undrained volumetric thermal expansion coefficient, in other words these parameters can be measured from tests where the fluid mass content is maintained constant.

1.3.8.3 Equation governing the variation in elementary Eulerian porosity

The Lagrangian and Eulerian porosities are related by the Jacobian of the transformation: $\Phi = J\phi$. Assuming infinitesimal transformations, we can easily obtain the following relation:

$$\Phi = (1+\varepsilon)\phi$$

A first order approximation yields:

$$d\Phi = d\phi + \Phi_o d\varepsilon$$

The constitutive equation (1.3.29) then leads to the equation governing the variation in elementary Eulerian porosity ϕ:

$$d\phi = (b - \Phi_o) d\varepsilon + \frac{dp_p}{N} - 3\alpha_\Phi dT \tag{1.3.36}$$

1.3.9 Nonlinear poroelastic behaviour

We will now adopt Biot's semilinear model (1973) which assumes that the nonlinear behaviour of some porous media is due to local geometry changes (variation in contact surfaces, closure of cracks, etc.) and that the behaviour of the solid matrix remains linear. We will also consider an isotropic porous medium and isothermal transformations. In addition, the reference state is assumed to be free of strains and pressure.

The semilinear constitutive equations are deduced by decomposing the stress state into an effective part acting on the "dry" medium (no fluid) and an isotropic part acting on the solid matrix [Bemer *et al.*, 2001]:

$$\underline{\underline{\sigma}} = \underline{\underline{\bar{\sigma}}} - p_p \underline{\underline{1}} \qquad (1.3.37)$$

$$p_p = 0 + p_p$$

Note that the effective stress $\underline{\underline{\bar{\sigma}}}$ defined in this way corresponds to Terzaghi's effective stress (Figure 1.3.2).

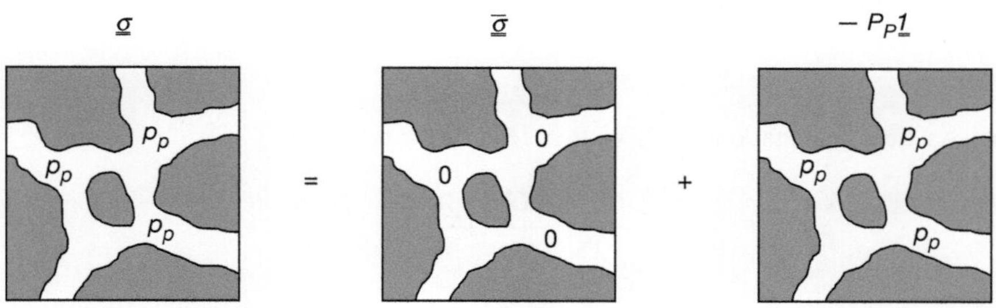

Figure 1.3.2

Load decomposition.

Since the matrix behaviour is assumed to be linear and isotropic with a bulk modulus K_s, decomposition of the stresses (1.3.37) is associated with the following decomposition of the strains:

$$\underline{\underline{\varepsilon}} = \underline{\underline{\bar{\varepsilon}}} - \frac{p_p}{3K_s} \underline{\underline{1}} \qquad (1.3.38)$$

where $\underline{\underline{\bar{\varepsilon}}}$ represents the strains produced by $\underline{\underline{\bar{\sigma}}}$.

Figure 1.3.3 illustrates the validity of Biot's semilinearity assumption for a 19.8% porosity reservoir sandstone. This experimental curve is obtained from a hydrostatic test where the confining pressure and the pore pressure are increased simultaneously, thereby avoiding any variation in the effective stress and the associated nonlinear strain.

Since only the strains $\underline{\underline{\bar{\varepsilon}}}$ include nonlinear variations, potential G_s is taken to be the sum of the quadratic potential of the linear poroelasticity G_s^{el} leading to the constitutive equations (1.3.29) and a third-order nonlinear potential H which is purely a function of $\underline{\underline{\bar{\varepsilon}}}$:

$$G_s\left(\underline{\underline{\varepsilon}}, p_p\right) = G_s^{el}\left(\underline{\underline{\varepsilon}}, p_p\right) + H\left(\underline{\underline{\bar{\varepsilon}}}\right) \qquad (1.3.39)$$

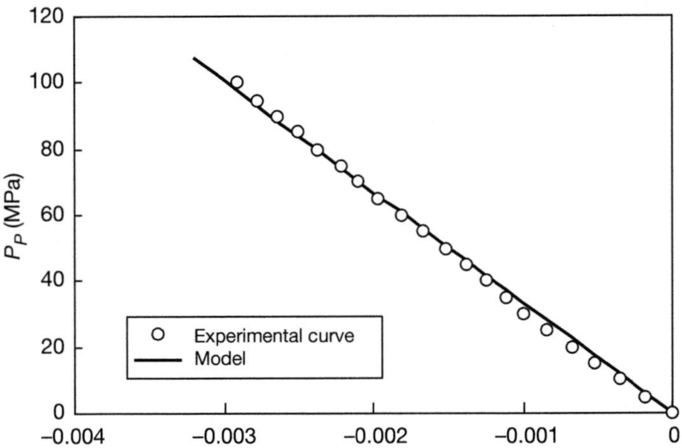

Figure 1.3.3

Illustration of the linear behaviour of the matrix.

The constitutive equations (1.3.26) then lead to the relations:

$$\underline{\underline{\sigma}} = \frac{\partial G_s^{el}}{\partial \underline{\underline{\varepsilon}}} + \frac{\partial H}{\partial \underline{\underline{\varepsilon}}}$$

$$\Phi = -\frac{\partial G_s^{el}}{\partial p_p} - \frac{1}{3K_s}\,\mathrm{tr}\frac{\partial H}{\partial \underline{\underline{\varepsilon}}}$$

Hence,

$$\begin{aligned} \underline{\underline{\sigma}} &= \left(K_d - \frac{2G}{3}\right)\varepsilon\,\underline{\underline{1}} + 2G\,\underline{\underline{\varepsilon}} - b\,p_p\,\underline{\underline{1}} + \frac{\partial H}{\partial \underline{\underline{\varepsilon}}} \\ \Phi - \Phi_o &= b\,\varepsilon + \frac{p_p}{N} - \frac{1}{3K_s}\,\mathrm{tr}\frac{\partial H}{\partial \underline{\underline{\varepsilon}}} \end{aligned} \qquad (1.3.40)$$

Equations (1.3.40) can be decomposed into a hydrostatic part and a deviatoric part:

$$\begin{aligned} \sigma &= K_d\,\varepsilon - b\,p_p + \frac{1}{3}\,\mathrm{tr}\frac{\partial H}{\partial \underline{\underline{\varepsilon}}} \\ \underline{\underline{s}} &= 2G\,\underline{\underline{e}} + \frac{\partial H}{\partial \underline{\underline{\varepsilon}}} - \frac{1}{3}\,\mathrm{tr}\frac{\partial H}{\partial \underline{\underline{\varepsilon}}} \\ \Phi - \Phi_o &= b\,\varepsilon + \frac{p_p}{N} - \frac{1}{3K_s}\,\mathrm{tr}\frac{\partial H}{\partial \underline{\underline{\varepsilon}}} \end{aligned} \qquad (1.3.41)$$

where $\sigma = \frac{\mathrm{tr}\,\underline{\underline{\sigma}}}{3}$, $\underline{\underline{s}} = \underline{\underline{\sigma}} - \sigma \underline{\underline{1}}$ and $\underline{\underline{e}} = \underline{\underline{\varepsilon}} - \frac{\varepsilon}{3}\underline{\underline{1}}$. Note that $\overline{\underline{\underline{s}}} = \underline{\underline{s}}$ and $\overline{\underline{\underline{e}}} = \underline{\underline{e}}$.

Biot (1963) adopts for H the following third-order expression:

$$H = D\frac{\overline{\varepsilon}^3}{3} + (F - D)\left(\overline{\varepsilon}\,\overline{I}_2 - 3\overline{I}_3\right) \qquad (1.3.42)$$

where $I_2 = \varepsilon_{11}\varepsilon_{22} + \varepsilon_{22}\varepsilon_{33} + \varepsilon_{33}\varepsilon_{11} - \varepsilon_{12}\varepsilon_{21} - \varepsilon_{23}\varepsilon_{32} - \varepsilon_{31}\varepsilon_{13}$ and $I_3 = \det(\varepsilon)$. F and D are constants which have the same dimensions as stress.

Taking the principal stress coordinate system, we obtain the constitutive equations:

$$\begin{aligned}
&\sigma = K_d\,\varepsilon - b\,p_p + \frac{D+2F}{3}\overline{\varepsilon}^2\\
&s_{11} = 2Ge_{11} + \frac{F-D}{3}\left[\overline{\varepsilon}_{22}^{\,2} + 2\overline{\varepsilon}_{11}\overline{\varepsilon}_{22} + \overline{\varepsilon}_{33}^{\,2} + 2\overline{\varepsilon}_{11}\overline{\varepsilon}_{33} - 2\left(\overline{\varepsilon}_{11}^{\,2} + 2\overline{\varepsilon}_{22}\overline{\varepsilon}_{33}\right)\right] \qquad (1.3.43a, b, c)\\
&\Phi - \Phi_o = b\,\varepsilon + \frac{p_p}{N} - \frac{D+2F}{3K_s}\overline{\varepsilon}^2
\end{aligned}$$

The expressions associated with the other principal stress directions are obtained simply by circular permutation.

After differentiating equations (1.3.43a) and (1.3.43c) we can explicitly define a tangent drained bulk modulus, a tangent Biot's coefficient and a tangent pore compressibility function of $\overline{\varepsilon}$. We obtain:

$$\begin{aligned}
d\sigma &= K_d^t\left(\overline{\varepsilon}\right)d\varepsilon - b^t\left(\overline{\varepsilon}\right)dp_p\\
d\Phi &= b^t\left(\overline{\varepsilon}\right)d\varepsilon + \left(\frac{1}{N}\right)^t\left(\overline{\varepsilon}\right)dp_p
\end{aligned} \qquad (1.3.44a, b)$$

with

$$\begin{aligned}
K_d^t\left(\overline{\varepsilon}\right) &= K_d + 2\frac{D+2F}{3}\overline{\varepsilon}\\
b^t\left(\overline{\varepsilon}\right) &= b - 2\frac{D+2F}{3K_s}\overline{\varepsilon}\\
\left(\frac{1}{N}\right)^t\left(\overline{\varepsilon}\right) &= \frac{1}{N} - \frac{D+2F}{3K_s^2}\overline{\varepsilon}
\end{aligned} \qquad (1.3.45a, b, c)$$

Note that the following relations are satisfied: $b^t = 1 - \frac{K_d^t}{K_s}$ and $\left(\frac{1}{N}\right)^t = \frac{b^t - \Phi_o}{K_s}$.

Equation (1.3.43a) can be used to express $\bar{\varepsilon}$ as a function of the mean effective stress $\bar{\sigma} = \sigma + p_p$ and therefore relate the tangent drained bulk modulus, the tangent Biot's coefficient and the tangent pore compressibility to the effective mean stress alone:

$$K_d^t(\bar{\sigma}) = \sqrt{K_d^2 + 4\frac{D+2F}{3}\bar{\sigma}}$$
$$b^t(\bar{\sigma}) = 1 - \sqrt{(1-b)^2 - 4\frac{D+2F}{3K_s}\frac{\bar{\sigma}}{K_s}} \quad (1.3.46a, b, c)$$
$$\left(\frac{1}{N}\right)^t = \frac{1-\Phi_o}{K_s} - \sqrt{\left(\frac{1-b}{K_s}\right)^2 - 4\frac{D+2F}{3K_s^2}\frac{\bar{\sigma}}{K_s^2}}$$

Figure 1.3.4 shows that the tangent drained bulk modulus does in fact vary as a function of the effective means stress $\bar{\sigma} = \sigma + p_p$.

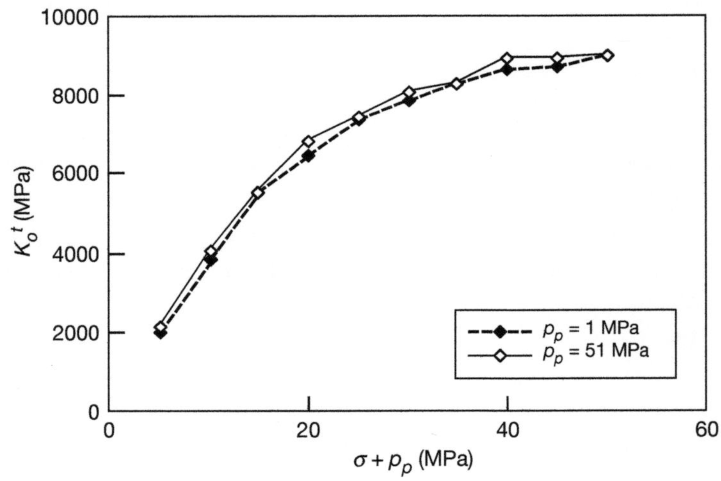

Figure 1.3.4

Variation law of the tangent drained bulk modulus on a sandstone with 19% porosity [Bemer *et al.*, 2001].

It is more difficult to define a tangent shear modulus using Biot's semilinear model. From the deviatoric part of the constitutive equations (1.3.43b), we must obtain a relation of type

$$d\underline{\underline{s}} = 2G^t\left(\underline{\underline{\varepsilon}}\right)d\underline{\underline{e}} \quad (1.3.47)$$

i.e., according to the principle direction 1,

$$ds_{11} = 2G^t\left(\underline{\underline{\varepsilon}}\right)de_{11} \quad (1.3.48)$$

However, by differentiating equation (1.3.43b), we obtain:

$$ds_{11} = 2G\,de_{11}$$
$$+2\frac{F-D}{3}\left[\left(-2\bar{\varepsilon}_{11}+\bar{\varepsilon}_{22}+\bar{\varepsilon}_{33}\right)d\bar{\varepsilon}_{11}+\left(\bar{\varepsilon}_{11}+\bar{\varepsilon}_{22}-2\bar{\varepsilon}_{33}\right)d\bar{\varepsilon}_{22}+\left(\bar{\varepsilon}_{11}-2\bar{\varepsilon}_{22}+\bar{\varepsilon}_{33}\right)d\bar{\varepsilon}_{33}\right]$$

To obtain an expression of type (1.3.48), relation $\bar{\varepsilon}_{33} = \bar{\varepsilon}_{22}$ must necessarily be satisfied, which amounts to having a condition of axisymmetry around principal axis 1. In this special case, equation (1.3.43b) leads to relation:

$$q = 3G\,\varepsilon_d + \frac{9}{4}(F-D)\varepsilon_d^2 \tag{1.3.49}$$

where $q = \sigma_{22} - \sigma_{11}$ is the deviatoric stress and $\varepsilon_d = \frac{2}{3}\left(\varepsilon_{22} - \varepsilon_{11}\right) = \bar{\varepsilon}_d$ the associated deviatoric strain.

A tangent shear modulus function of ε_d can be defined explicitly by differentiating equation (1.3.49). We obtain:

$$dq = 3G^t\left(\varepsilon_d\right)d\varepsilon_d \tag{1.3.50}$$

with
$$G^t\left(\varepsilon_d\right) = G + 3\frac{F-D}{2}\varepsilon_d \tag{1.3.51}$$

Equation (1.3.49) can be used to express ε_d as a function of the deviatoric stress q and thereby relate the tangent shear modulus to the deviatoric stress alone:

$$G^t\left(q\right) = \sqrt{G^2 + \left(F-D\right)q} \tag{1.3.52}$$

To date, the validity of this variation law has not been checked experimentally.

1.3.10 Poroelastoplastic behaviour

The poroelastic model described here is the modified Cam-Clay model [Roscoe and Burland, 1968]. This model was developed to represent the isothermal behaviour of normally consolidated or slightly overconsolidated clays. It is an isotropic poroelastoplastic behaviour model with isotropic strain hardening, defined for a water-saturated porous medium and therefore initially associated with a model describing the viscous flow of a low compressibility fluid. Extending poroelastoplasticity to unsaturated porous media results in relatively complex behaviour models. Some examples can be found in [Alonso *et al.*, 1990] and [Dangla *et al.*, 1997].

1.3.10.1 Plastic strains, plastic porosity and trapped energy

Apart from interstitial pressure and strain-hardening parameter(s), the yield function of an isotropic material depends on the first three invariants of the stress tensor. Models often implement yield functions which depend only on the first two invariants. The mean stress p and the deviatoric stress q are generally used in rock mechanics [Charlez, 1994]:

$$p=\frac{tr\,\underline{\underline{\sigma}}}{3} \text{ and } q=\sqrt{\frac{3}{2}\underline{\underline{s}}:\underline{\underline{s}}} \text{ with } \underline{\underline{s}}=\underline{\underline{\sigma}}-\frac{tr\,\underline{\underline{\sigma}}}{3}\underline{\underline{1}}$$

The volume strain ε_v and the deviatoric strain ε_d correspond respectively to the mean and deviatoric stresses:

$$\varepsilon_v=tr\,\underline{\underline{\varepsilon}} \text{ and } \varepsilon_d=\sqrt{\frac{2}{3}\underline{\underline{e}}:\underline{\underline{e}}} \text{ with } \underline{\underline{e}}=\underline{\underline{\varepsilon}}-\frac{tr\,\underline{\underline{\varepsilon}}}{3}\underline{\underline{1}}$$

Stresses p and q are the variables associated with strains ε_v and ε_d by the density of free energy ψ_s. We obtain:

$$\underline{\underline{\sigma}}:d\underline{\underline{\varepsilon}}=p\,d\varepsilon_v+q\,d\varepsilon_d$$

Since the intrinsic dissipation is positive (1.3.16), we obtain the inequality:

$$\varphi_s=p\frac{d\varepsilon_v}{dt}+q\frac{d\varepsilon_d}{dt}+p_p\frac{d\Phi}{dt}-\frac{d\psi_s}{dt}\geq 0 \tag{1.3.53}$$

When considering evolutions where the internal variables are constant, the following constitutive equations can be deduced from (1.3.53):

$$p=\frac{\partial\psi_s}{\partial\varepsilon_v} \qquad q=\frac{\partial\psi_s}{\partial\varepsilon_d} \qquad p_p=\frac{\partial\psi_s}{\partial\Phi} \tag{1.3.54}$$

Coussy defines poroelastoplasticity as being "*the ability of porous materials to undergo permanent strains and permanent changes in porosity and, as a consequence, permanent changes in fluid mass content*" [Coussy, 2004]. The strain increments can thus be decomposed into an elastic part and a plastic part:

$$d\varepsilon_v=d\varepsilon_v^e+d\varepsilon_v^p \text{ and } d\varepsilon_d=d\varepsilon_d^e+d\varepsilon_d^p \tag{1.3.55}$$

Rather than a plastic variation in fluid mass content, it is generally preferred to introduce a plastic Lagrangian porosity variation [Dangla *et al.*, 1997]. We then write:

$$d\Phi=d\Phi^e+d\Phi^p \tag{1.3.56}$$

A thermodynamic state is characterised by the external state variables: ε_v, ε_d and Φ, the internal state variables: ε_v^p, ε_d^p, Φ^p and variables characterising the strain hardening $\underline{\chi}$:

$$\psi_s=\psi_s\left(\varepsilon_v,\varepsilon_d,\Phi,\varepsilon_v^p,\varepsilon_d^p,\Phi^p,\underline{\chi}\right)$$

Using the energy separation assumption, we can express the free energy of the skeleton in the following form [Coussy, 2004]:

$$\psi_s=\psi_s^{el}\left(\varepsilon_v-\varepsilon_v^p,\varepsilon_d-\varepsilon_d^p,\Phi-\Phi^p\right)+U\left(\underline{\chi}\right) \tag{1.3.57}$$

where ψ_s^{el} represents the proportion of the free energy of the skeleton recoverable as mechanical energy and therefore depends only on reversible strains and porosities. U represents the quantity of energy trapped through the strain-hardening phenomena and is therefore related to the strain-hardening variables.

By introducing equations (1.3.54) and (1.3.57) in inequality (1.3.53), we obtain the following intrinsic dissipation expression:

$$\varphi_s = p\,\frac{d\varepsilon_v^p}{dt} + q\,\frac{d\varepsilon_d^p}{dt} + p_p\,\frac{d\Phi^p}{dt} - \frac{\partial U}{\partial \underline{\chi}} \cdot \frac{d\underline{\chi}}{dt} \geq 0 \qquad (1.3.58)$$

which demonstrates that the thermodynamic forces associated with ε_v^p, ε_d^p, Φ^p and $\underline{\chi}$, are respectively p, q, p_p, and the thermodynamic force $\underline{\varsigma}$ defined by the relation:

$$\underline{\varsigma} = -\frac{\partial U}{\partial \chi} \qquad (1.3.59)$$

1.3.10.2 Cam-Clay model constitutive equations

Using the same quadratic expression adopted in the context of linear poroelasticity for ψ_s^{el}, we can deduce from equations (1.3.54) and (1.3.57) the following expression of the constitutive equations:

$$\begin{aligned} dp' &= d(p + dp_p) = K_0(d\varepsilon_v - d\varepsilon_v^p) \\ dq &= 3G(d\varepsilon_d - d\varepsilon_v^p) \\ d\Phi - d\Phi^p &= b(d\varepsilon_v - d\varepsilon_v^p) + \frac{dp_p}{N} \end{aligned} \qquad (1.3.60\text{a, b, c})$$

where $p' = p + b\,p_p$ represents the elastic mean effective stress.

For a low compressibility interstitial fluid, equation (1.3.60c) can also be expressed as a function of the fluid mass content m:

$$\frac{dm}{\rho_o^{fl}} - d\Phi^p = -b\left(d\varepsilon_v - d\varepsilon_v{}^p\right) + \frac{dp}{M} \qquad (1.3.61)$$

The constitutive equations (1.3.60) define a model whose elastic part is linear. The model defined by Roscoe, however, assumes that the constitutive equations are in fact nonlinear. This special formulation is based on Terzaghi's hypothesis, which assumes elastic and plastic incompressibility of the solid matrix. The elastic and plastic effective stresses are then identical to Terzaghi's effective stresses, defined by: $\underline{\underline{\sigma}}' = \underline{\underline{\sigma}} + p\underline{\underline{1}}$ [Charlez, 1994].

The choice of a nonlinear constitutive equation is based on experimental observations which show that, under hydrostatic loading, the void ratio decreases in direct proportion to the logarithm of the absolute value of the mean effective stress. Hence,

$$de^e = -\kappa\,\frac{d\,p'}{p'} \qquad (1.3.62)$$

where κ is the swelling coefficient. For small strains and an incompressible matrix, the void ratio is related to the volume strain by $d\varepsilon_v = de/1+e_o$. The nonlinear volumetric constitutive equation of the modified Cam-Clay model is then written:

$$d\varepsilon_v - d\varepsilon_v^p = -\frac{\kappa}{1+e_o}\frac{dp'}{p'} \tag{1.3.63}$$

The deviatoric constitutive equation is assumed to be linear, thus:

$$d\varepsilon_d - d\varepsilon_d^p = \frac{dq}{3G} \tag{1.3.64}$$

The model formulated in this way is thermodynamically admissible [Coussy, 2004].

In view of the various constitutive equations defined above, two formulations of the Cam-Clay model can be considered:

- a linear formulation defined by equations (1.3.60a, b),
- a semilinear formulation defined by equations (1.3.63) and (1.3.64).

Unlike the linear model, the semilinear model is based on Terzaghi's hypothesis. Note that constitutive equation (1.3.60a) is obtained by linearising relation (1.3.63) around a mean effective stress p_o'.

1.3.10.3 Plasticity criterion

The plasticity criterion of the Cam-Clay model is defined by the following yield function:

$$f\left(p, q, p_p, p_{cr}\right) = \frac{1}{2}\left[\left(p + \beta p_p + p_{cr}\right)^2 + \frac{q^2}{\mathcal{M}^2} - p_{cr}^{\ 2}\right]$$

where β and M are constants and p_{cr} is a positive strain-hardening variable called the (current plastic) critical effective pressure which is associated with the scalar thermodynamic force ς.

$p_\beta' = p + \beta p_p$ is the plastic mean effective stress, not to be confused with the elastic mean effective stress $p' = p + bp_p$. The yield surface f is then written:

$$f\left(p_\beta', q, p_{cr}\right) = \frac{1}{2}\left[\left(p_\beta' + p_{cr}\right)^2 + \frac{q^2}{\mathcal{M}^2} - p_{cr}^{\ 2}\right] \tag{1.3.65}$$

In plane $\left(p_\beta', q\right)$, the plasticity criterion $f\left(p_\beta', q, p_{cr}\right) = 0$ defines a family of ellipses passing through the origin, having a maximum value at $p_\beta' = -p_{cr}$ and cutting the x-axis at $p_\beta' = -2p_{cr}$ (Figure 1.3.5). The latter point, which corresponds to the current elastic limit of the material under hydrostatic loading, is called the effective consolidation pressure. Double

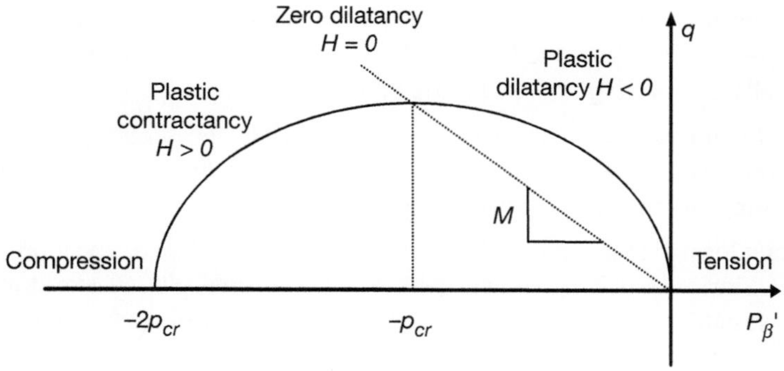

Figure 1.3.5

Cam-Clay model yield surface.

the value of the critical pressure represents in fact the absolute value of the maximum plastic mean effective pressure previously undergone by the material. A material is therefore said to be normally consolidated (or overconsolidated) if the current plastic mean effective stress is equal to (or less than in absolute value) the consolidation effective pressure.

1.3.10.4 Flow rule

The flow rule defines the evolution of plastic strains when the plasticity criterion has been reached. The Cam-Clay model considers a normal flow rule associated with the plasticity criterion:

$$d\varepsilon_v^p = d\lambda \frac{\partial f}{\partial p} \qquad d\varepsilon_d^p = d\lambda \frac{\partial f}{\partial q} \qquad d\Phi^p = d\lambda \frac{\partial f}{\partial p_p}$$

Hence,

$$d\varepsilon_v^p = d\lambda \left(p_\beta' + p_{cr} \right)$$

$$d\varepsilon_d^p = d\lambda \frac{q}{\mathcal{M}^2} \qquad (1.3.66)$$

$$d\Phi^p = d\lambda \, \beta \left(p_\beta' + p_{cr} \right)$$

with $d\lambda = \frac{1}{H}\left[\left(p_\beta' + p_{cr} \right) dp_\beta' + \frac{q\,dq}{\mathcal{M}^2} \right] \geq 0$, where H is the strain-hardening modulus.

From the above expressions, we deduce that $d\Phi^p = \beta\, d\varepsilon_v^p$.

We will now consider the straight line of equation $p_\beta' + p_{cr} = 0$. This straight line is a line of symmetry for the yield surface which it separates into two equal semi-ellipses.

For points on the yield surface belonging to the semi-ellipse located in the half-plane defined by $p_\beta{}'+p_{cr}<0$, flow rule (1.3.66) shows that $d\varepsilon_v^p<0$ and $d\Phi^p<0$. The material is then plastically compacting and the plastic porosity decreases.

Inversely, for points on the yield surface belonging to the semi-ellipse located in the half-plane defined by $p_\beta{}'+p_{cr}>0$, we have $d\varepsilon_v^p>0$ and $d\Phi^p>0$. The material is then plastically dilating and the plastic porosity increases.

For the point located at the intersection between the straight line of equation $p_\beta{}'+p_{cr}=0$ and the ellipse, flow rule (1.3.66) shows that $d\varepsilon_v^p=0$ and $d\Phi^p=0$. Plastic strain then takes place with no variation in volume and the plastic porosity remains unchanged.

1.3.10.5 Strain-hardening equations

The Cam-Clay model introduces a potential $h\left(p,p_p,p_{cr}\right)$ not associated with the criterion to specify the evolution equations governing the scalar hardening variable χ:

$$h\left(p,p_p,p_{cr}\right)=\frac{1}{2}\frac{\beta-\Phi_o}{\left(1-\Phi_o\right)^2}\left(p+\beta p_p+p_{cr}\right)^2 \tag{1.3.67}$$

χ is then related to the critical pressure p_{cr} by:

$$d\chi=d\lambda\frac{\partial h}{\partial p_{cr}}=d\lambda\frac{\beta-\Phi_o}{\left(1-\Phi_o\right)^2}\left(p+\beta p_p+p_{cr}\right) \tag{1.3.68}$$

Relations (1.3.66) and (1.3.68) show that variable χ can be identified with the plastic void ratio e^p, which represents the irreversible variation of void ratio with respect to a given initial reference state:

$$e^p=\frac{\Phi^p-\Phi_o\varepsilon_v^p}{\left(1-\Phi_o\right)^2} \tag{1.3.69}$$

The expression of p_{cr} as a function of e^p remains to be determined. Experience suggests using an equation of type [Charlez, 1994] (Figure 3.1.6):

$$de^p=-\left(\lambda-\kappa\right)\frac{d\,p_{cr}}{p_{cr}}\quad\text{with }\lambda>\kappa \tag{1.3.70}$$

where λ is the (plastic) compressibility coefficient.

Relation (1.3.70) can equally be expressed as a function of the plastic volume strain:

$$\frac{d\,p_{cr}}{p_{cr}}=-\frac{1}{\lambda-\kappa}\frac{\beta-\Phi_o}{\left(1-\Phi_o\right)^2}d\varepsilon_v^p \tag{1.3.71}$$

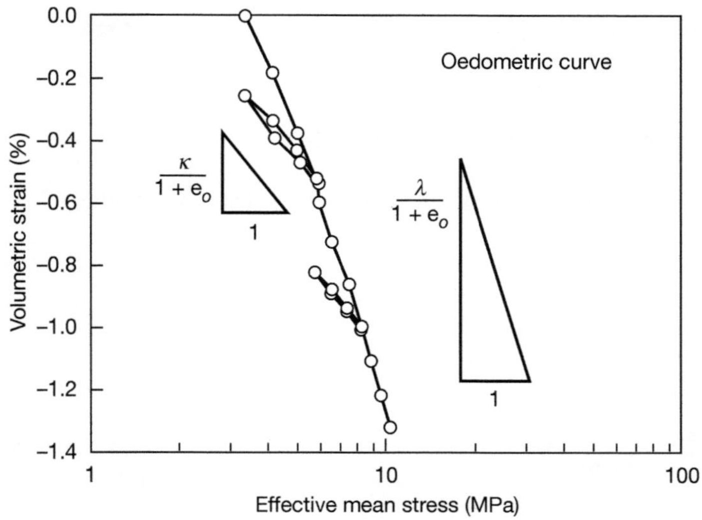

Figure 1.3.6

Poroplastic behaviour of an argillite in an oedometric test (under Terzaghi's hypothesis).

Relations (1.3.59) and (1.3.70) lead to the following expression of trapped energy:

$$U\left(e^p\right)=\left(\lambda-\kappa\right)p^o_{cr}\exp\left[-\frac{e^p}{\lambda-\kappa}\right] \tag{1.3.72}$$

The strain-hardening modulus H is then:

$$H=\frac{1}{\lambda-\kappa}\frac{\beta-\Phi_o}{\left(1-\Phi_o\right)^2}p_{cr}\left(p+\beta p_p+p_{cr}\right)\left(p+\beta p_p\right) \tag{1.3.73}$$

Consequently, for yield points such that the material exhibits plastically compacting behaviour, the hardening modulus is positive and the critical pressure increases, which corresponds to a strain-hardening phase. Inversely, for yield points such that the material has a plastically dilating behaviour, the hardening modulus is negative and the critical pressure decreases, which corresponds to a strain-softening phase. For yield points such that the plastic strain takes place with no variation in volume or plastic porosity, the hardening modulus is zero and the material behaves as a plastically perfect material. For these points, the strains which are now only plastic occur at constant stresses and interstitial pressure. The state is said to be critical.

The locus of critical states is defined by the following relations:

$$p+\beta p_p+p^o_{cr}\exp\left[-\frac{e^p}{\lambda-\kappa}\right]=0 \qquad q-\mathcal{M}\,p^o_{cr}\exp\left[-\frac{e^p}{\lambda-\kappa}\right]=0$$

In plane $\left(p_\beta',q\right)$, it is the straight line of equation $q = -\mathcal{M}\,p_\beta$ (Figure 1.3.5).

1.3.10.6 Incremental expression of the flow rule

Introducing expression (1.3.73) of the hardening modulus in flow rule (1.3.66) leads to the following incremental relations:

$$d\varepsilon_v^p = \frac{\left(1-\Phi_o\right)^2}{\beta-\Phi_o}(\lambda-\kappa)\left[\left(\frac{1}{p_{cr}}+\frac{1}{p_\beta'}\right)dp_\beta'+\frac{q}{\mathcal{M}^2\,p_{cr}\,p_\beta'}dq\right]$$

$$d\varepsilon_d^p = \frac{\left(1-\Phi_o\right)^2}{\beta-\Phi_o}(\lambda-\kappa)\left[\frac{q}{\mathcal{M}^2\,p_{cr}\,p_\beta'}dp_\beta'+\frac{q^2}{\mathcal{M}^4\,p_{cr}\,p_\beta'\left(p_{cr}+p_\beta'\right)}dq\right] \qquad (1.3.74)$$

$$d\phi^p = \beta\,d\varepsilon_v^p$$

Eliminating p_{cr} by using relation $f\left(p_\beta',q,p_{cr}\right)=0$ and introducing parameter $\eta = q/p_\beta'$, we finally obtain:

$$d\varepsilon_v^p = \frac{\left(1-\Phi_o\right)^2}{\beta-\Phi_o}(\lambda-\kappa)\left[-\frac{dp_\beta'}{p_\beta'}-\frac{2\eta\,d\eta}{\mathcal{M}^2+\eta^2}\right]$$

$$d\varepsilon_d^p = \frac{\left(1-\Phi_o\right)^2}{\beta-\Phi_o}(\lambda-\kappa)\frac{2\eta}{\mathcal{M}^2-\eta^2}\left[-\frac{dp_\beta'}{p_\beta'}-\frac{2\eta\,d\eta}{\mathcal{M}^2+\eta^2}\right] \qquad (1.3.75)$$

$$d\phi^p = \beta\,d\varepsilon_v^p$$

1.4 DETERMINATION OF *IN SITU* STRESSES

It is impossible to measure the stress tensor directly. However, some components can be estimated by observing the strain processes associated with the action of the stresses on the material considered.

In the absence of tectonic effects, the major principal stress is assumed to be the vertical stress σ_v due to the weight of the overburden. The two other principal stresses are horizontal and equal to $K\sigma_v$. According to the theory of elasticity, if we assume that there is no horizontal strain during burial (oedometric condition) and that the medium is isotropic and homogeneous, then we obtain:

$\sigma_H' = \dfrac{\nu_d}{1-\nu_d}\sigma_v'$, where σ' are effective stresses and ν_d the drained Poisson's ratio.

In most cases, the *in situ* stress state does not correspond to this ideal scheme and the stress distribution is more complex, depending on geological history. We identify:

- regional stresses which are related to major geological phenomena (plate tectonics); indications of their orientation and sometimes their amplitudes are now available [Reinecker *et al.*, 2005],
- local stresses which are affected by local accidents such as folds, faults, salt domes; stress rotations may then lead to orientations and amplitudes differing considerably from the original scheme.

In foothill regions, for example, tectonic forces dominate and the maximum horizontal stress may be the highest of the principal stresses [Last *et al.*, 1995].

1.4.1 Determination from *in situ* measurements

We assume that the lithostatic vertical stress is one of the principal stresses.

1.4.1.1 Horizontal stress directions

Two methods can be used to determine the *in situ* horizontal stress directions in vertical boreholes:

- The first is based on analysis of borehole breakouts. Breakouts form at different depths when the concentration of compressive stresses exceeds the rock strength. The axis defined by breakout corresponds to the minimum horizontal stress azimuth.
- The second method uses the traces of drilling-induced hydraulic fractures (tensile failures). Their azimuth is a reliable indicator of the maximum horizontal stress directions (Figure 1.4.1).

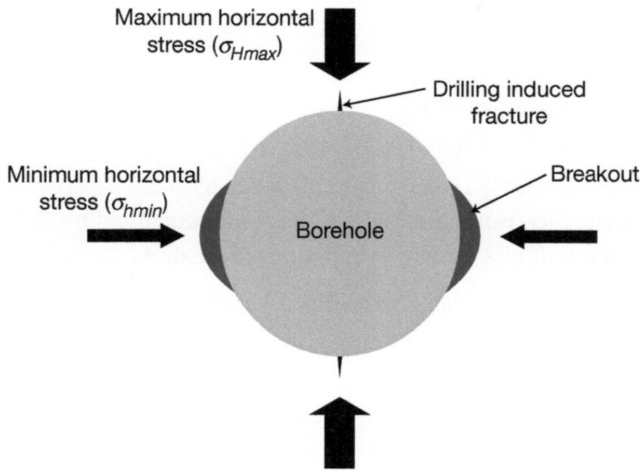

Figure 1.4.1

Breakouts and fractures induced in the stress field.

Breakouts and drilling-induced fractures can be detected by systems recording the hole geometry, multi-arm callipers and well imaging probes.

1.4.1.1.1 Oriented calliper data (HDT, SHDT, HEXDIP)

HDT (High resolution Dipmeter Tool) and SHDT (Stratigraphic High resolution Dipmeter Tool) are Schlumberger tools with four hydraulically operated orthogonal pads. Each pad is equipped with a microresistivity electrode (1 for the HDT, 2 for the SHDT).

The HEXDIP (Hexagonal Dipmeter) manufactured by Baker has six pads.

The 4-pad calliper data are composed of:

- calliper C1-3: diameter between pad 1 and pad 3,
- calliper C2-4: diameter between pad 2 and pad 4,
- P1 azimuth: bearing of pad 1 from the north,
- relative bearing: bearing of the top side of the tool from the north,
- hole azimuth: direction of the borehole in a horizontal plane from the north,
- deviation: hole deviation from the vertical.

As it is brought up to the surface, the calliper logging tool is rotated constantly. In zones with breakouts, the pads tend to remain blocked and the tool bearing remains frozen (Figure 1.4.2).

Figure 1.4.3 shows an example of a calliper and the common interpretation in terms of hole geometry which is made in terms of in-gauge hole, with breakout, eroded with formation of washout, deformed on one side only (keyseat) by the drillstring [Plumb & Hickman, 1985].

The criteria for accurate identification of breakout are:

- the borehole deviation must be less than 10°,
- the log must show tool rotation above and below the breakout,
- the tool rotation in the zone concerned must be less than 5°,
- the difference between the two callipers must be more than 1 cm,
- the zone must be more than 3 m long.

1.4.1.1.2 Imaging data

Acoustic imaging is extremely useful for identification of breakouts that appear as dark bands on opposite sides of a wellbore wall (Figure 1.4.4):

- UBI (Ultrasonic Borehole Imager) manufactured by Schlumberger,
- CBIL (Circumference Borehole Imaging Log) manufactured by Baker Atlas,
- CAST (Circumferential Acoustic Scanning Tool) manufactured by Halliburton.

Resistive imaging also produces good results for identification of breakouts and drilling-induced fractures (Figure 1.4.5):

- FMS (Formation MicroScanner), FMI (Formation MicroImager), OBMI (Oil-Base MicroImager) manufactured by Schlumberger. The FMI has a network of 24 electrodes on each of its 4 orthogonal pads [Ma *et al.*, 1993].
- STAR (Simultaneous Acoustic and Resistivity) manufactured by Baker Atlas includes acoustic and resistive imaging sensors in a single instrument. The resistive probe is equipped with a network of 24 electrodes on each of its six pads.

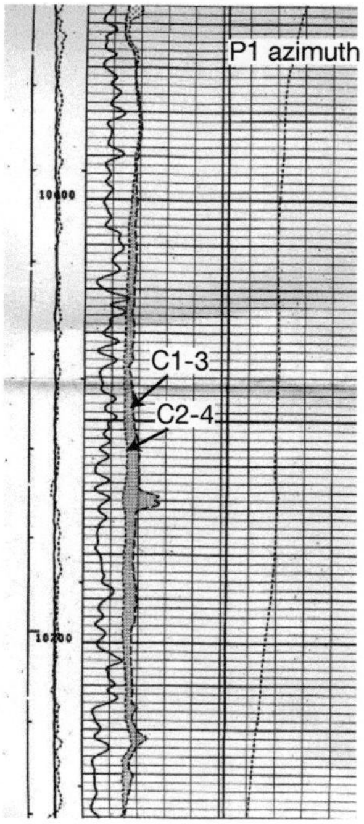

Figure 1.4.2

Example of breakout detection.

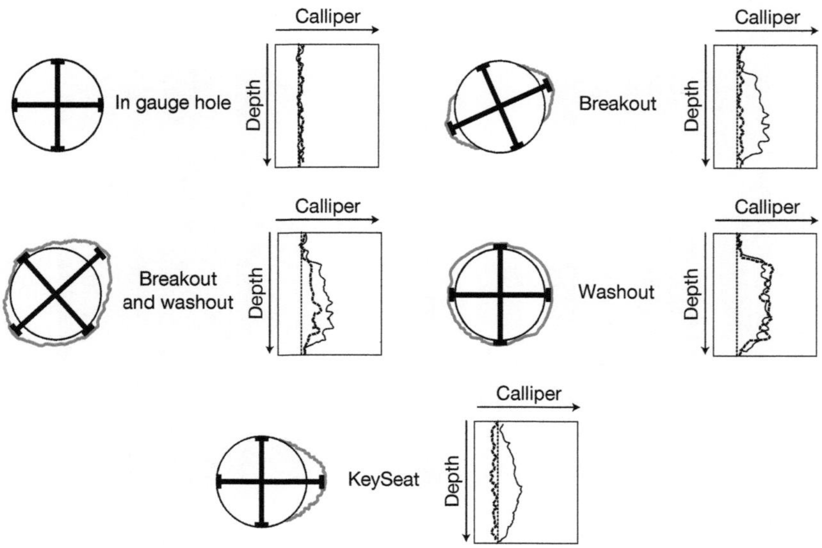

Figure 1.4.3

Interpretation of CALLIPER data [Plumb & Hickman, 1985].

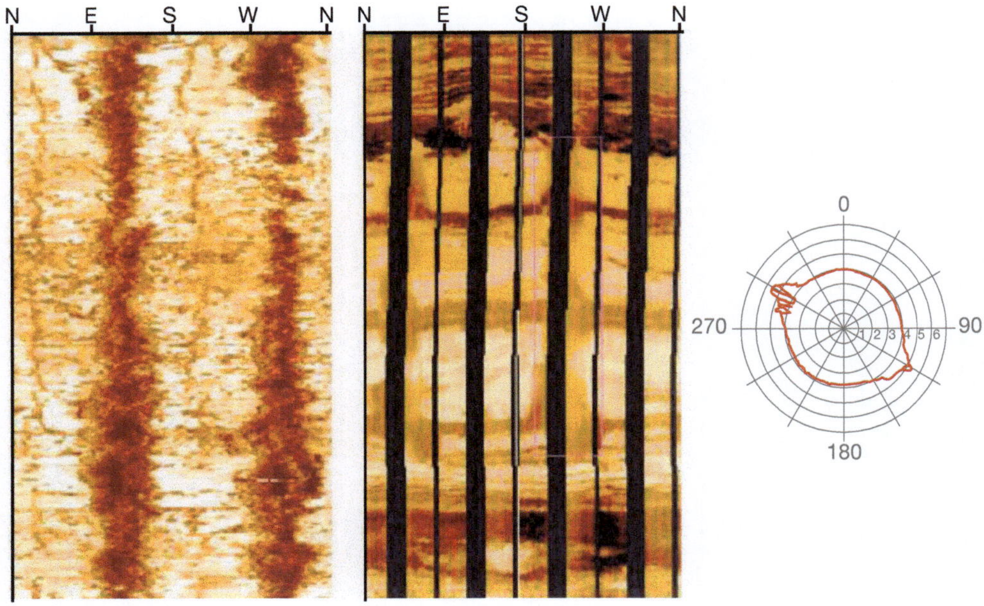

Figure 1.4.4

Interpretation of UBI (a) and FMI (b) images in terms of breakouts [Zoback *et al.*, 2003].

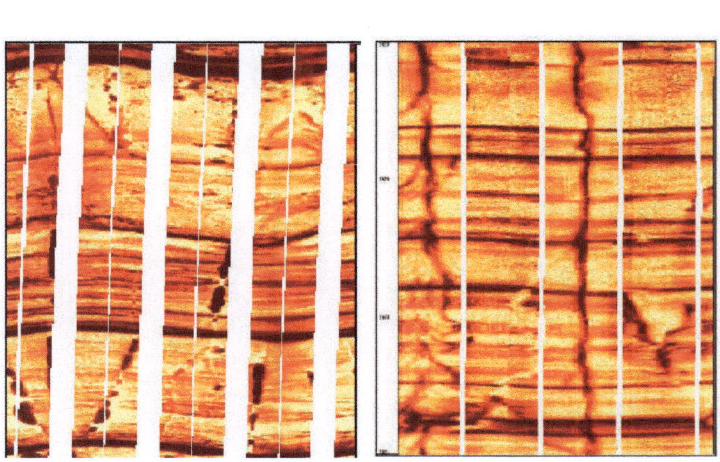

Figure 1.4.5

FMI images interpreted in terms of drilling-induced fractures. These fractures are characterised by low vertical extension and irregular shape and can only appear on one side of the hole (HEF petrophysical consulting site).

1.4.1.1.3 Shear wave anisotropy

Anisotropy of shear wave propagation velocities may indicate stress anisotropy, which must not be confused with structural anisotropy [Plona *et al.*, 2000]. Fast shear azimuth is assumed to coincide with the direction of maximum horizontal stresses. The latest generation tools (MSIP, sonic scanner) are even very accurate [Pistre *et al.*, 2005] (Figure 1.4.6).

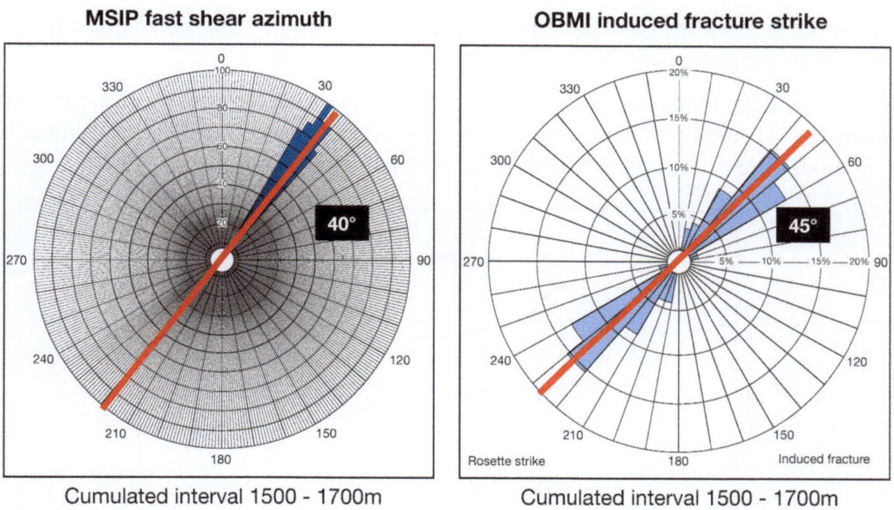

Figure 1.4.6

Comparison between MSIP fast shear azimuth and OBMI induced fractures [Franco *et al.*, 2005].

1.4.1.2 Stress amplitude

The amplitude of *in situ* stresses is generally determined by assuming that the principal stresses are vertical and horizontal. Starting from this assumption, the analysis is carried out to obtain the vertical stress on the basis of a density log, assuming that the vertical stress is equal to the weight of overburden. The horizontal stresses are obtained from well tests (Mini-Frac tests, LOT, ELOT, SRT).

The stress tensor measurement obtained is local information; the distribution at field scale can be disturbed by faults, fractures and heterogeneities of mechanical properties.

1.4.1.2.1 Determining the amplitude of vertical stresses

The vertical stress can only be determined by integrating the density log. In this case, several problems may occur:

- availability of density logs over a reduced thickness,
- error when interpreting the values,
- possible arching effect in an anticline, increasing horizontal stresses and decreasing vertical stresses.

In offshore, the density log can be calibrated on measurements taken in geotechnical surveys performed before installation of platforms or subsea pipelines.

1.4.1.2.2 Determining the amplitude of horizontal stresses

Amplitude of the minimum horizontal stress

The minimum stress amplitude can only be measured directly by using hydraulic fracturing tests. Fractures propagate perpendicular to the minimum stress.

In the hydraulic fracturing test, a section of the borehole, less than one metre long, is sealed off by two packers. The sealed-off section is then slowly pressurised with a fluid, water or mud. Figure 1.4.7 illustrates the results obtained.

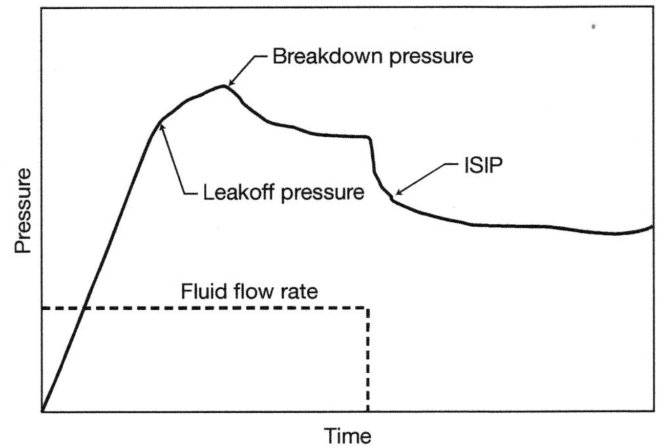

Figure 1.4.7

Hydraulic fracturing test.

The pressure is increased by injecting a fluid at constant flow rate. The pressure curve can be subdivided into five phases: a linear increase phase, a nonlinear phase during which the rate of increase slows down, the maximum pressure (breakdown pressure), a more or less steady decrease indicating the opening of a fracture, lastly the relaxation curve after the injection pump stops. The Leak-Off Pressure (LOP) corresponds to the end of the linear section with change of slope. During the relaxation phase, the Instantaneous Shut In Pressure (ISIP) or closure pressure is assimilated to the minimum stress *in situ*. The shut-in pressure corresponds to the level of pressure exerted by the fluid inside the fracture, just sufficient to keep the edges of the fracture open. Consequently, the minimum stress amplitude can be determined by analysing the pressure during fracturing and closure of the fracture. Several methods are available for determining closure pressure from the fall off curve. One of these uses a change in linearity of the pressure decay on a graph of pressure plotted against the square root of time [Desroches and Kurkjian, 1998, Zoback, 2008]. If the amplitude of the closure pressure is less than the overburden at this depth, then the smallest principal stress will correspond to the horizontal minimum stress (S_{hmin}).

The Leak-Off Test (LOT) is conducted during the drilling operations after cementing a casing. It consists in increasing the pressure at the casing shoe to a value greater than the LOP but less than the rock breakdown pressure (Figure 1.4.8). This operation is mainly used to evaluate the maximum permissible mud pressure. In a variant of the LOT, the Extended Leak-Off Test (ELOT), the pressurisation phase continues beyond the formation breakdown pressure and the fracture propagates. The test consists of two or more pressurisation cycles (Figure 1.4.9). Extended leak-off tests use procedures from leak-off tests, therefore open bottomhole or "barefoot" well configuration is pressurised without using packers.

In practice, the quality of 50% of the LOTs is poor. There are several reasons for this: open section too long, mud compressibility artefact, pumping stopped too soon or relaxation phase badly recorded [Chardac *et al.*, 2005]. LOT and ISIP generally overestimate the minimum stress.

A good estimate of the minimum stress amplitude can generally be obtained from a series of fracturing tests. ELOTs are reliable tests compared with LOTs, but they are not as precise as mini-frac tests with several pressurisation cycles.

After several pressurisation cycles, when injection is stopped, fluid is quickly withdrawn from the fracture to close it in the vicinity of the wellbore only. The remainder of the fracture stays pressurised above the closure pressure. A pressure rebound can then be observed when the fracture produces back until pressure decreases again. If the maximum pressure reached during flow back is greater than the hydrostatic pressure, this is a good indication that the fracture has been created [Desroches and Kurkjian, 1999] and, for an impermeable formation, this leads to an estimation of the minimum horizontal stress.

SRTs (Step Rate Tests) are carried out to determine the maximum water injection rate possible without fracturing a permeable formation. The test is conducted by recording the injection pressure and injection rate at several increasing injection rates, allowing the

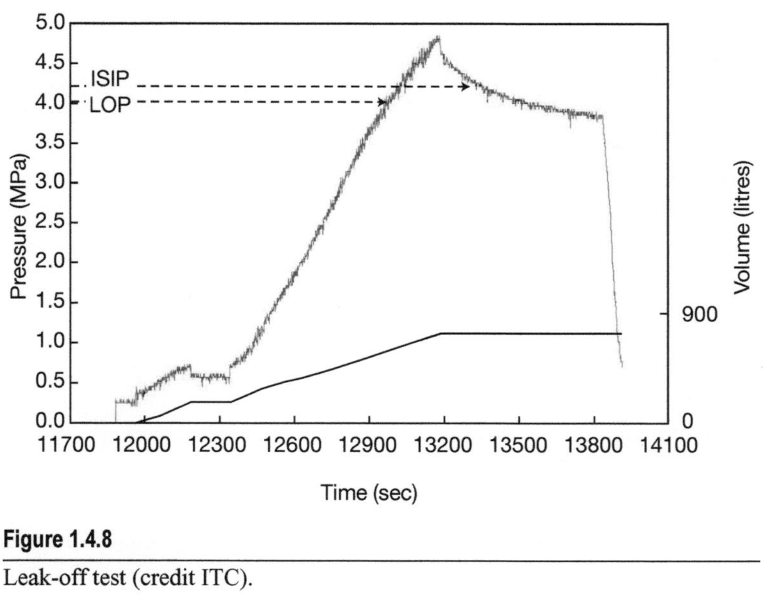

Figure 1.4.8

Leak-off test (credit ITC).

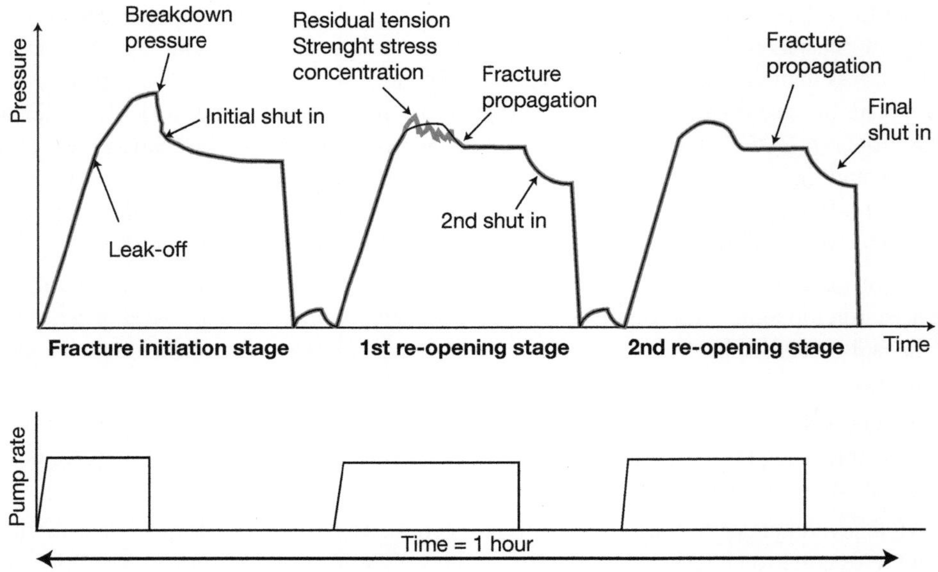

Figure 1.4.9

Typical ELOT result [Enever *et al.*, 1996].

pressure to stabilise at each step. All steps must last the same time, for example 30 min. The fracture pressure can be identified from a graph of injection pressures against flow rates. The SRTs obviously give an upper limit of the minimum horizontal stress.

In all cases an approach with a "reconciliation" diagram between all the tests performed is necessary [Desroches and Kurkjian, 1999].

1.4.1.2.3 Maximum horizontal stress

Currently, no method is available to measure the maximum horizontal stress. In some specific cases an approximation can be found, especially if the vertical stress is the major stress.

We have the relation:

$$Pb - Pp = \frac{(3\sigma_{Hmin} - \sigma_{Hmax} + Tc - 2Pp)}{K}$$

with

- Pb: breakdown pressure,
- Pp: pore pressure,
- Tc: tensile strength,
- K: poro-elastic constant which depends on the permeability and compressibility of the formation,
- $K = 2$ is applicable in permeable formations in which the injected fluid penetrates the porous space during the test.

$K = 1$ is applicable in impermeable formations where the pore pressure remains unchanged in the porous space around the borehole [Lowrey & Steiner, 1993].

When the stress $S_{h\,min}$ is known, the fracture reopening pressure can be used to determine the maximum stress. This pressure is expressed as: $P_{reopening} = 3S_{h\,min} - S_{H\,max} - P_p$ (K = 1 and Tc = 0).

When the rock strength is known, the breakout angular width [Zobak *et al.*, 2003] measured on acoustic image log can be used.

$$S_{H\,max} = \frac{(C + 2P_p + \Delta P + \sigma^{\Delta T}) - S_{h\,min}(1 + 2\cos 2\theta_b)}{1 - 2\cos 2\theta_b}$$

with

$2\theta_b = \pi - w_{bo}$ with w_{bo} angular breakout opening,
C:rock cohesion,
Pp:pore pressure,
$\sigma^{\Delta T}$:thermal stress,
ΔP:difference between the well pressure during drilling and the pore pressure.

The measurements of shear wave velocities made with crossed dipole sonic logging tools such as DSI, MSIP, Sonic scanner, Wavesonic (fast and slow D_t shear) supply anisotropy indicators. These indicators are expressed as percentages. For example, *DT*-based anisotropy is:

$$\frac{Dt_{slow} * - Dt_{fast}}{(Dt_{slow} + Dt_{fast})/2} * 100$$

Calibration on mini-frac values or poroelastic models is necessary to convert these indicators in terms of minimum and maximum stress amplitudes [Klimentos *et al.*, 1999, Plona *et al.*, 2000, Sinha *et al.*, 2000]. Anisotropies as low as 1% can be measured [Franco *et al.*, 2005].

If LOT data are available in several inclined boreholes (with different inclinations and azimuths) distributed in a homogeneous field in a specific zone, an inversion method can be used to estimate the orientation and amplitude of the horizontal stresses [Aadnoy, 1988].

1.4.2 Determination from core measurements

After extraction, a core expands more in the direction of the maximum stresses. This expansion generates microcracks perpendicular to the three principal stresses. Two methods are based on this relaxation principle:

- ASR (Anelastic Strain Recovery),
- DSCA (Differential Strain Curve Analysis).

With these methods, the core must be oriented, which is often costly, and highly accurate strain measurements are required. With ASR, the measurements must be taken immediately after recovery since anelastic strains develop in less than 50 hours. A detailed interpretation of the tests is provided in the book by Fjaer *et al.* (1991). DSCA requires isotropic reconfinement. Although inaccurate, these two methods provide an idea of the orientation, and even the amplitude, of the stresses [Cui *et al.*, 2006]. The measurements become very difficult to interpret in anisotropic and/or fractured rocks.

1.4.3 How to constrain the stress tensor

Geological strata contain faults, fractures and discontinuities at several scales. Friction in these existing discontinuities limits the stresses.

At equilibrium, Jaeger and Cook demonstrated that the ratio of the maximum and minimum effective stresses had to remain less than a certain value which depends on the coefficient of friction of the discontinuity μ:

$$\frac{\sigma'_1}{\sigma'_3} \leq \left[\left(\mu^2+1\right)^{\frac{1}{2}}+\mu\right]^2$$

The values of σ'_1 and σ'_3 depend on the tectonic system:

- normal fault stress regime ($\sigma_v > \sigma_{H\,max} > \sigma_{h\,min}$): $\sigma'_1 = \sigma'_v$ and $\sigma'_3 = \sigma'_{h\,min}$,
- strike-slip fault stress regime ($\sigma_{H\,max} > \sigma_v > \sigma_{h\,min}$): $\sigma'_1 = \sigma'_{H\,max}$ and $\sigma'3 = \sigma'_{h\,min}$,
- reverse stress regime ($\sigma_{H\,max} > \sigma_{h\,min} > \sigma_v$): $\sigma'_1 = \sigma'_{H\,max}$, $\sigma'_3 = \sigma'_v$.

According to Byerlee [1978], the coefficient of friction of the discontinuities μ lies between 0.6 and 1.

Geomechanical modelling tools can be used to constrain the stress tensor, checking the stability or instability of the boreholes where clearly identified breakouts and induced fractures appear.

These tools require the following input data:

- the assumed directions and amplitudes of the minimum and maximum horizontal stresses,
- the pore pressure,
- the geomechanical properties of the rocks where the breakouts appear,
- a failure criterion (e.g. Mohr Coulomb).

Indications such as mud losses during drilling may also be used.

1.5 GEOMECHANICAL MODELLING ELEMENTS

The discipline of oil reservoir geomechanical modelling is in full expansion. The first studies were conducted to explain the significant subsidence observed in some mature fields with highly porous (North Sea chalk) or poorly consolidated (Zuata field, Venezuela) reservoirs [Charlez, 1997]. The scope of geomechanical modelling is much broader and concerns all stress-dependent problems: modification of the mud window, instability of the borehole walls, modification of flow directions, reactivation of fractures or faults, interpretation of modifications in 4D seismic data, microseismicity. The stress modifications are induced by reservoir depletion or, on the contrary, injection of fluids during hydrocarbon production or gas storage (natural or CO_2).

1.5.1 Geology and griding

The zone considered for geomechanical modelling must include a volume larger than the reservoir since the mechanical properties of surrounding formations have an impact on the mechanical behaviour of the reservoir. The width of the sideburden will be ideally 3 times the horizontal extension of the reservoir region. The height of the underburden will be of the same order of value as the depth of the reservoir region.

Geomechanical modelling may be more or less complex, 2D or 3D depending on the problem to be solved. The geometry of the various layers is derived from seismic interpretation, constrained by the available borehole data. The number of stratigraphic units to be taken into account depends on the borehole data available, especially the seismic impedance contrast (product of the formation density by the compression wave velocity). The dynamic elastic properties from the sonic logs and the static elastic properties used in the geomechanical models are in fact closely related.

The mesh will be oriented according to the orientation of the principal stresses.

The geological strata exhibit discontinuities which may be more or less detectable and more or less cemented or open. They share the following characteristics: low shear strength, negligible tensile strength and hydraulic conductivity often much greater than the wall rock. Several categories can be identified:

- diaclases: rock fractures, due to compressive, tensile or shear failures related to the tectonic stresses; the two parts of the rock have not moved, however;
- faults: fractures identical to the diaclases but which caused a relative movement of two parts of the wall rock. A slip has therefore occurred along this fault and the throw varies from a few tens of centimetres to a few hundred metres;
- sedimentary joints: these are the joints separating two strata deposited at different times under different conditions.

Discontinuities are generally areas of high deformability, where failure is easier and where the fluids can flow.

All realistic models of a structure and its environment must take into account the possible discontinuities. Geometric discontinuities are generally correctly taken into account.

1.5.2 Geomechanical properties

Geomechanical properties must be attributed at least to each rock layer identified in the model. Seismic, log analysis and laboratory data can be used.

Resistivity, nuclear and gamma ray logs give a good estimation of the mineralogy and, in particular, the location of the shale layers [Chardac *et al.*, 2005].

The compression and shear wave velocities (*Vp* and *Vs* if available) and the density logs provide information on the dynamic elastic properties, but the plasticity or viscosity properties can only be obtained by conducting laboratory tests.

A few calibration points can be obtained from the laboratory data, for the conversion of sonic logs into static elastic properties.

If no S-wave records are available, the *Vs* velocity is estimated from Vp using empirical formulae that depend on the lithology [Castagna *et al.*, 1985].

The *Vp* and *Vs* velocities are used to calculate the undrained dynamic moduli:

$$K_u^{dyn} = \rho\left(V_p^2 - \frac{4}{3}V_s^2\right) \text{ and } G^{dyn} = \rho V_s^2, \text{ with}$$

K_u^{dyn}: rock undrained dynamic bulk modulus

G^{dyn}: rock dynamic shear modulus

ρ: overall rock density (deduced for example from the density log)

$$E_u^{dyn} = \rho V_S^2 \frac{\left(3V_P^2 - 4V_S^2\right)}{\left(V_P^2 - V_S^2\right)} \text{ and } \nu_u^{dyn} = \frac{\left(V_P^2 - 2V_S^2\right)}{2\left(V_P^2 - V_S^2\right)}$$

The static properties can be deduced from the dynamic properties [Vidal-Gilbert *et al.*, 2005]. The procedure consists of two phases. Firstly, "fluid saturation" correction is carried out to change from undrained dynamic elastic moduli to dynamic drained moduli. Secondly, dynamic drained moduli are changed to static drained moduli.

One of the problems frequently studied in rock physics, especially for analysis of logs, samples and seismic data, is the prediction of seismic velocities in rocks saturated with a given fluid from another saturation state. This fluid substitution problem can be predicted using the Biot-Gassmann theory (1951). Gassmann calculated, for a quasi static load, the difference in elastic modulus between a drained porous medium and the same medium in undrained state, in which local variations in fluid content are nil during mechanical loading. The Gassmann equation can be expressed as:

$$\frac{K_u}{K_s - K_u} = \frac{K_d}{K_s - K_d} + \frac{K_{fl}}{\phi\left(K_s - K_{fl}\right)}$$

with

K_u: undrained bulk modulus

K_d: drained bulk modulus

K_s: bulk modulus of the matrix (mineral bulk material forming the rock)

K_{fl}: fluid bulk modulus

The following assumptions are used in the Biot-Gassmann equation:

1) the porous medium is macroscopically homogeneous and isotropic,
2) all pores are connected,
3) the pores are filled with a low-viscosity fluid,
4) the solid-fluid system is closed (undrained),
5) as the wave passes, there is no coupling (relative movement) between the solid phase and the liquid phase,
6) the fluid does not interact with the solid, so that the shear modulus remains constant.

A practical expression of the Biot-Gassmann equation involves the Biot coefficient b and the Biot modulus M:

$$K_u = K_d + b^2 M$$

with:

$$b = 1 - \left(\frac{K_d}{K_s}\right)$$

and

$$\frac{1}{M} = \frac{b-\phi}{K_s} + \frac{\phi}{K_{fl}}$$

In this expression, we can see that the first term in the expression of 1/M is often much smaller that the second since K_s is always much bigger than K_{fl}. Neglecting this first term, we obtain a very simple expression:

$$K_u \approx K_d + \frac{b^2}{\phi} K_{fl},$$ [Rasolofosaon and Zinszner, 2003].

The error thus committed has been estimated to be less than 5%.

In order to apply the Gassmann equation to wave propagation (dynamic case), the frequency must be low enough for the variation in saturating fluid induced by passage of the seismic wave to be uniform throughout the porous space (no gradient). This low frequency limitation explains why the Biot-Gassmann equation is well suited to seismic frequencies (< 100 Hz). While this relation can also be applied for the log analysis frequency band, it may prove less relevant for ultrasound measurements on samples. The latter point is open to discussion since Rasolofosaon and Zinszner (2004) demonstrated that the experimental results on samples seemed to be in agreement with the Biot-Gassmann theory, for various fluids saturating the porous medium.

The Biot-Gassman equation is then frequently used to estimate the dynamic drained moduli from undrained dynamic moduli:

$$K_d^{dyn} = K_u^{dyn} - \frac{K_{fl}\left(1-\frac{K_u^{dyn}}{K_s}\right)^2}{\left(1-\frac{K_{fl}}{K_s}\right)\phi - \frac{K_{fl}}{K_s}\left(1-\frac{K_u^{dyn}}{K_s}\right)} \quad \text{with } K_u^{dyn} = \frac{E_u^{dyn}}{3\left(1-2\nu_u\right)} \quad \text{where:}$$

K_s: bulk modulus of the solid matrix (of the material forming the grains),
K_{fl}: bulk modulus of the fluid saturating the pores,
ϕ: porosity (deduced from the density measurements, for example).

The bulk modulus of the rock matrix K_s depends on the mineral composition, but also to the second order on pressure and temperature. To a first approximation, the orders of magnitude indicated in Table 1.5.2 (dynamic values) will be used.

Table 1.5.2 Mean values of dynamic bulk and shear moduli in the main minerals forming sedimentary rocks [Zinszner and Pellerin, 2007].

Mineral class	ρ(kg/m^3)	K_s(GPa)	G_s(GPa)
Clay estimation	2,750	25	9
Quartz	2,650	37	45
Calcite	2,710	70	30
Dolomite	2,870	80	50
Siderite	3,960	120	50

If the mineral composition is known, homogenisation methods can be applied to estimate the mean effective modulus of the matrix, or at least determine its upper and lower limits [Berryman *et al.*, 1995, Hu *et al.*, 2006]. This is rarely the case outside the reservoir.

The fluid bulk modulus K_{fl} depends on the nature of the saturating fluid. For a water-oil mixture for example, it can be calculated using the following formula:

$$K_{fl} = \frac{1}{\frac{S_w}{K_w} + \frac{(1-S_w)}{K_{oil}}}$$

with:

K_w: water bulk modulus,
K_{oil}: oil bulk modulus,
S_w: water saturation, determined using log analyses, for example.

The water bulk modulus varies with salinity, pressure and temperature (Figure 1.5.1). For brine at a concentration of 15 g/L, $K_{fl} = 2.2667 + 0.0063$ p in GPa with p = pressure expressed in MPa [Zinszner and Rasolofosaon, 2008].

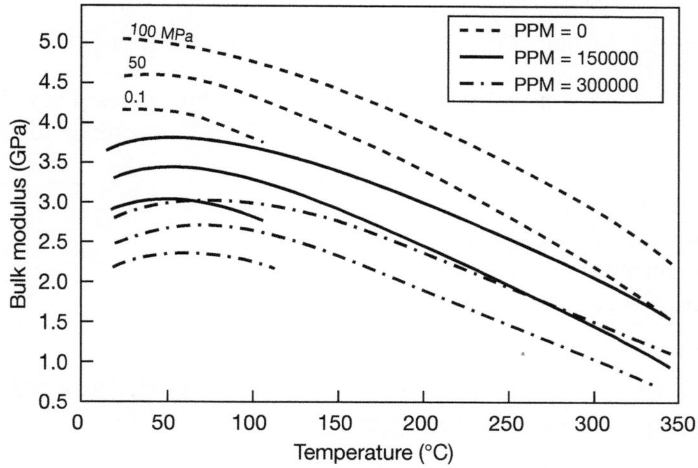

Figure 1.5.1

Bulk modulus of a brine against pressure, temperature and salinity [Batzle and Wang, 1992].

The bulk modulus of oil depends on its density, as well as the pressure and temperature (Figure 1.5.2).

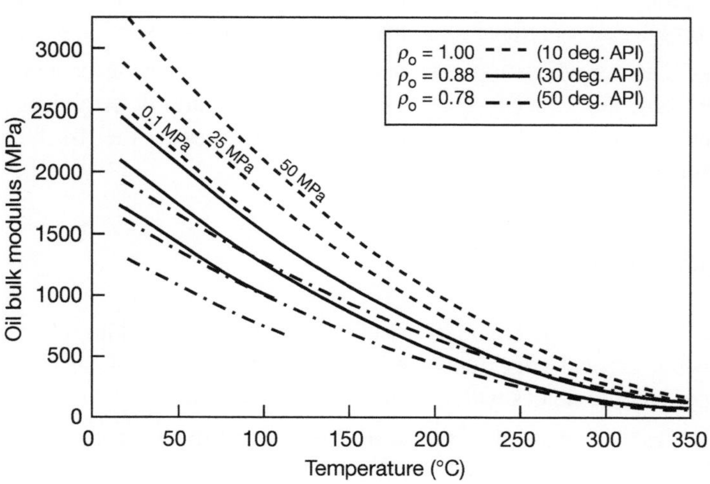

Figure 1.5.2

Bulk modulus of oil against pressure, temperature and composition [Batzle and Wang, 1992].

The estimated dynamic drained moduli can then be changed to static drained moduli using empirical correlations [Wang, 2000].

For reservoir rocks, we can use the correlations determined by Wang, who identifies two types:

- if $E_d^{stat} < 15$ GPa then $E_d^{stat} = 0.4145 E_d^{dyn} - 1.0595\,(GPa)$ and
- if $E_d^{stat} > 15$ GPa, then $E_d^{stat} = 1.1530 E_d^{dyn} - 15.1970\,(GPa)$.

For shales, we can use correlations determined for all lithologies [Lacy, 1997, Cooper and Hatherly, 2003] or for shales only [Horsrud, 2001], according to the following sequences:

- *Vp Vs*: P and S wave velocities,
- *Ed*: dynamic modulus,
- *Es*: static modulus,
- *UCS*: unconfined compressive strength.

For example:

- $E_d = 0.265(V_p)^{2.04}$ [Lacy, 1997] with Vp in km/s and Ed in millions of psi (all lithologies),
- $UCS = A.Vp^3$ [Cooper & Hatherly, 2003],
- $E_S = 0.0428(E_d)^2 + 0.2334 E_d$ (for clays),
- $UCS = 0.2787(E_S)^2 + 2.458 E_S$ (all lithologies),

$E_s = 0.076(V_p)^{3,23}$ $G = 0.03(V_p)^{3.3}$ and $UCS = 0.77(Vp)^{2.93}$ (for shales) [Horsrud, 2001].

The static Young's moduli are lower than the dynamic moduli [Wang, 2000; Yale *et al.*, 1995]. The static Poisson's ratios are greater than or equal to the dynamic Poisson's ratios.

For both static and dynamic properties, laboratory measurements remain preferable when determining local or regional relations of the zone of interest.

Obviously no matter which approach is chosen, the measurements cannot be considered as completely reliable. Current geomechanical models are deterministic. The values assigned to the geomechanical properties must take into account the risk factor associated with the problem studied.

1.5.3 Boundary conditions

The traditional displacement boundary conditions consist in blocking the horizontal displacements of the vertical edges and the vertical displacements of the horizontal edges, unless the upper edge coincides with that of the natural terrain.

The vertical displacements are fixed at the bottom and the horizontal displacements on the lateral sides.

The contact surfaces between layers and the faults are generally fixed in a first simulation. Friction can then be introduced, or not, depending on the resulting stress state.

2

Geomechanics, Drilling and Production

2.1 DRILLING PERFORMANCE

Drilling bits can be divided into two broad categories: roller-cone bits and PDC (Polycrystalline Diamond Compact) bits (Figure 2.1.1).

Roller-cone bits have several toothed cones which are free to turn as the bit is rotating. Tri-cone is the dominant type. The teeth can either be ground in the matrix or consist of tungsten carbide buttons inserted into holes in the cones. The teeth are designed to crush the rock as the bit rotates (Figure 2.1.2).

PDC bits consist of several discs containing numerous cutting edges made of tungsten and diamond. They cut the rock by shearing, layer after layer. These bits, more modern than roller-cone bits, can reach high drilling rates in relatively soft formations. They are also more efficient in hard rocks, provided that certain recommendations are respected (speed and weight on bit), especially for new bits which are more fragile.

Drillers have two main concerns:

- reaching maximum efficiency under ordinary conditions,
- solving "problems" under exceptional conditions which could eventually damage the bit, in particular.

The "problems" most frequently encountered are excessive vibrations, whose effects at the surface arouse the driller's attention, although lateral vibrations and precession sometimes remain unnoticed.

These excessive vibrations are generally observed when drilling through hard rocks at low rate of penetration (regular bounce or precession), but the TRAFOR recordings taken by IFP Energies nouvelles have also demonstrated bounce activity (more chaotic than the previous activity), while drilling rapidly through soft rocks; expulsion of the cuttings in "plugs" seems to be the most likely cause for the excitation [Putot, 1995].

Until recent years in the drilling industry, "rock drillability" and "drillstring dynamics" were quite independent disciplines [Pessier and Fear, 1992].

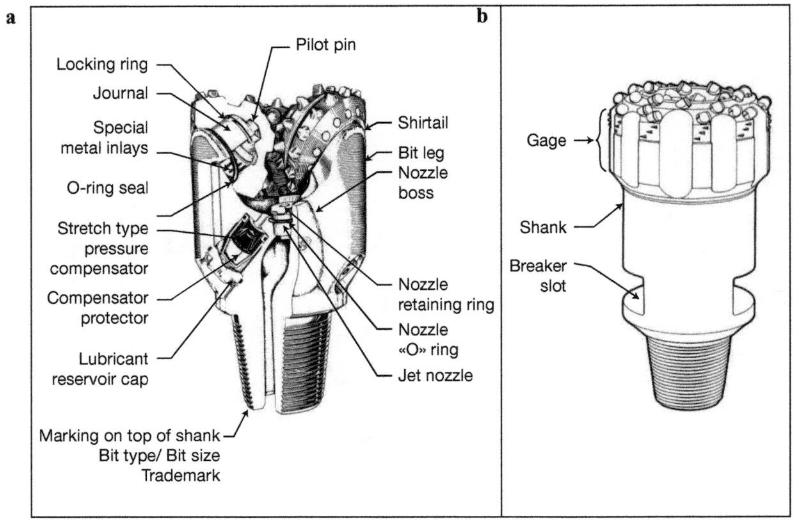

Figure 2.1.1

Examples of bits: roller-cone (a) and PDC (b) [Nguyen, J-P., 1993].

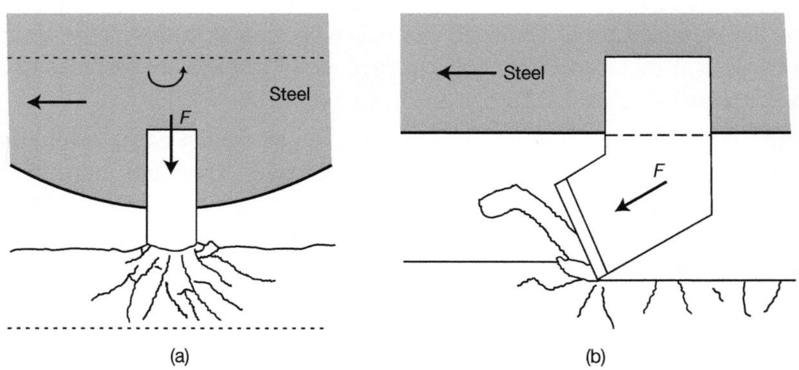

Figure 2.1.2

Rock destruction modes: punching for the roller-cone bit, shearing for the new PDC bit [Hélène Geoffroy, 1996].

"Rock drillability" concerns the ability of the drill bit to penetrate and remove the rock, in the context of the bottomhole, controlled from the surface via the drillstring. This speciality generally deals with ordinary drilling situations and aims at optimising bit concepts with respect to cut and hydraulics.

"Drillstring dynamics" studies examine the types of vibration which would damage the drilling system equipment (drillstring and bit), using in particular surface indicators more or less able to detect the various types of vibration at the bottom. This speciality also analyses the conditions leading to a vibratory malfunction.

With traditional rock drillability, it is impossible to correctly study the reduced drilling rate regimes with high dissipation at the cutting face and therefore validly assess the drilling efficiency [Warren and Smith, 1985]; the advance speed is reduced considerably by the vibrations; in addition, the dynamics of the bottomhole drilling system, in particular, is strongly related to the aggressivity of the cutters and the conditions under which the cuttings are expelled. The distinction between the two disciplines therefore becomes quite arbitrary. Drillability in dynamic regime is the end result of combining the two disciplines.

As regards drillability, one direction consists in analysing drillings in relatively soft formations. Tool performance, mainly that of monobloc "PDC" tools, is generally not a major problem, provided that the cuttings are suitably evacuated, which nevertheless demands well-designed tool hydraulics, good control and, in particular, sufficient hydraulic power [Wardlaw, 1971].

It seemed useful to describe the phenomenon associated with a tool cleaning fault, through balance equations both as regards the divided solid (rock) and the drilling mud, relying on simplified behaviour laws [Putot, 1995]. The evolution of the mixture of fluid and rock particles is governed by a "source" term including the conditions under which the rock is broken by the cutters, together with terms of retention on the particles crushed but not necessarily expelled, function of the entrainment capacity of the drilling mud.

Note that the hydraulics is a determining factor in soft rocks for evacuation of the cuttings but also essential in "hard" rocks for good cooling of the cutters, guarantee of their integrity.

2.1.1 Empirical drilling efficiency: weight on bit, rate of penetration and torque

2.1.1.1 Axial kinematics of the bit: normalised rate of penetration

The first means of analysing the drilling performance are modelled on the concerns of the driller who, as a priority, pays special attention to the axial kinematics of the bit.

Main notations (with variants):

W	Weight On Bit (WOB)
T	Torque On Bit (TOB)
R	Rate Of Penetration (ROP)
N	bit speed of rotation (RPM)
$\delta = R/N$	penetration per revolution

Penetration per revolution divided by weight on bit (δ/W) is an intuitive concept and a key indicator: a high value is a sign of good performance, while a low value often indicates an accumulation of cuttings and clogging.

Historically, the first refinement made to the driller's good judgement – do not drill too quickly or the hole will clog up – consisted in recording the *d*-exponent indicator:

$$d = \frac{\log R/N}{\log W/D} \tag{2.1.1}$$

where D is the bit diameter.

Under normal conditions, the *d*-exponent indicator can be used to "subtract" the effects of modifications of the "loading" W and N. Its "logarithmic" definition offers the advantage of slightly reducing the experimental dispersion, since the signal is generally very noisy due to fluctuations related to the mineralogical characteristics of the rocks at grain scale and to the pore pressure.

Plotted against the depth for a given lithology, the d-exponent provides an image of the compaction. In a normal situation, the greater the depth, the greater the compaction. It can be used to detect any compaction anomaly, for example areas of pore overpressure [Mouchet and Mitchell, 1989].

Real-time drilling interpretation is commonly based on observing the normalised rate of penetration R_D alone, which provides information on the lithology changes and would also give an indication on bit operation. The following expression was proposed by Warren (1984):

$$R_D = k\left(\frac{W}{\sigma D^2}\right)^n \qquad (2.1.2)$$

where:

σ is the unconfined compressive strength of the rock, frequently compared with a hardness.

k is an empirical proportionality factor related to the bit efficiency (ability to break the rock).

n is an exponent characterising, in principle, the local failure mechanism but which, in practice, is more representative of a degree of uncertainty related to the position on the curve $\delta = f(W)$ (Figure 2.1.3).

$n = 2$ is the value generally used to characterise field data; it corresponds to a normal drilling regime, for which increasing the weight on bit W significantly increases the performance, i.e. the penetration per revolution, $\delta = R/N$. We simply take the logarithm of relation (2.1.2) to obtain the d-exponent (2.1.1).

A decrease in n indicates *a priori* a deterioration of drilling conditions, generally due to difficult expulsion of rock cuttings and sometimes more or less sudden damage to the drill bit. The weight then has less effect on the normalised penetration δ. These approximations reflect the existence of an "S-shaped curve" whose upper plateau corresponds to saturation of δ with W (Figure 2.1.3).

For both roller-cone bit and PDC, Falconer *et al.* (1988) demonstrated that using the assumptions adopted in his model, R_D (ability of the bit to penetrate the rock) is inversely proportional to the shear strength τ of the rock; paradoxically, for the roller-cone bit, although the kinematics consists mainly of punching, the efficiency of the shear mechanism is the determining factor when assessing penetration. The paradox can be explained by the fact that the implicit assumption τ/σ = constant is often made.

Falconer *et al.* (1988) proposed a quantity *FORS* proportional to the drilling resistance defined later by Detournay and Defourny (1992) for the PDCs:

$$FORS = \frac{W}{R_D D^2} = \frac{\sigma}{k} \qquad (2.1.3)$$

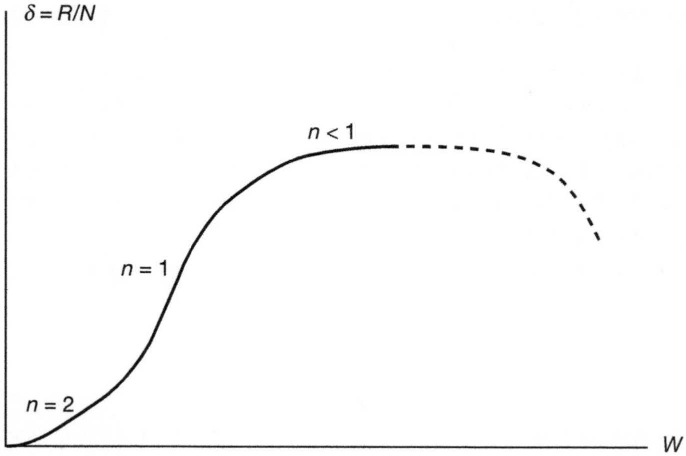

Figure 2.1.3

Sigmoid shape of the graph of penetration per revolution d against weight W: the exponent n reflects the local behaviour.

FORS is the ratio of rock penetration resistance σ divided by bit efficiency factor k.

Cunningham (1978) proposed a numerical frame very similar to expression (2.1.2), where the exponents assigned to the hardness σ and the weight W in the expression of their dependency with respect to the normalised penetration differ – which is not the case in (2.1.2) – and are fully defined by the rock hardness; he improves his description by explicitly introducing the drop in performance resulting from the pressure difference between mud pressure and pore pressure, using the following empirical relation where he assumes that the deviations from linearity $R(N)$ are due to effects of pressure difference Δp:

$$R = \frac{NW^{\mathrm{a}}}{k_1 \sigma_{\mathrm{d}}^{1.5} + \sqrt{NW^{\mathrm{a}}}\,(\Delta p)^{0.75}} \tag{2.1.4}$$

Numerous laboratory experiments have corroborated these expressions.

The previous literal expression, essentially empirical, showing the effect of interstitial pressure, explains the role of this pressure in the shape of the curves $R(N)$ at fixed W; it provides information on the decrease in penetration $\delta = R/N$ with speed N, useful to characterise the dependency of torque on bit with angular speed, especially to assess the stick-slip mechanism.

As an illustration, Cunningham observes:

- linearity $R(N)$ for $\Delta p = 0$,
- $R(N)$ in $N^{0.65}$ for $\Delta p = 20$ MPa (soft rock),
- $R(N)$ in $N^{0.85}$ for $\Delta p = 20$ MPa (hard rock).

For more recent studies concerning the effect of pore pressure on drilling performance, interested readers can refer in particular to Peltier and Atkinson (1986), Lesso and Burgess (1986), Detournay and Atkinson (1991), Hanson (1991).

2.1.1.2 Circumferential response and representation in the plane $R_D - T_D$

The second generation of performance indicators is no longer based on a single scalar quantity but on the definition of an operation line in the plane $R_D - T_D$, the energy component of the torque T_D forming a second indicator just as valuable as the axial component. This circumferential component indicates the shear strength, generally moderate for roller-cone bits and much higher for PDC type monobloc bits. It is the energy response of the torque, since the weight does not act directly in the drilling process (Figure 2.1.4).

The adimensional variable T_D introduced by par Falconer *et al.* (1988) can be written simply:

$$T_D = \frac{1}{D}\frac{T}{W} \tag{2.1.5}$$

It suggests a friction law (Amonton's 2nd law postulates that friction is proportional to the normal load). Using the Pessier and Fear (1992) approximations for a simplified bit model, we then obtain:

$$T_D = \frac{\mu}{3} \tag{2.1.6}$$

where μ is the "coefficient of friction".

T_D is in fact the transposition, at bit scale, of the ratio τ/σ shear strength divided by punching strength. This quantity indicates the drilling efficiency.

A reasonable value of μ for a rock-mechanical system interaction is about 0.6 [Scholz, 1990], so we can expect T_D to have values of about 0.2.

In practice, Falconer records values in the region of 0.1 for a roller-cone bit operating normally, in marls, and between 0.20 and 0.35 for a new PDC bit; equivalent friction, so the performances of the two types of bit are therefore *a priori* quite different, at least when new. These differences can be explained by the fact that the rock destruction modes are radically different: punching for the roller-cone bit (little friction), shear for the PDC (high interaction).

Values of T_D significantly higher than the previous reference values generally indicate *porous* formations (e.g. some sandstones) if, at the same time, R_D is relatively high. Inversely, significantly smaller values of T_D indicate stiff rock layers: indurated sandstones, compact limestones, dolomites, etc. provided that they are correlated with relatively low values of R_D.

The mechanical characterisations of the behaviour of roller-cone bits have been formulated by Falconer *et al.* (1988) and have been represented analytically in the plane of variables R_D and T_D.

The authors' analysis is based mainly on the mechanics of a circular cone interacting, on a contact surface close to a generator, with a flat block. The model had been presented previously by Warren (1984, 1985), with a proposal for extension to PDC bits made by Falconer without making any special adaptations to the theory.

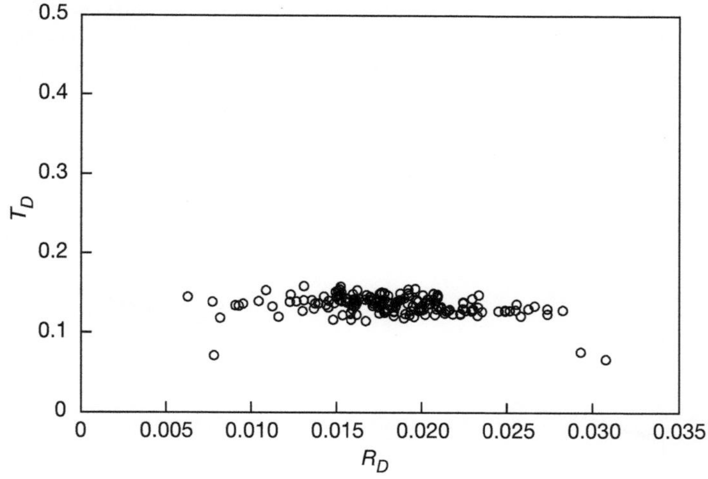

Figure 2.1.4

Example of $T_D - R_D$ values in a shale layer at a depth of approximately 1 800 m.

Detournay and Defourny (1992) completely revised the formulation for PDC bits by introducing assumptions on the local mechanics of the cutters and by including the forces on the entire tool. To characterise the results, variables S and E quite similar to the previous one are used, called respectively *drilling resistance* and *specific energy*. Detournay and Defourny also characterise the operation of roller-cone bits in the same system of variables, if only to clearly identify their specificity.

Note that, for PDC bits, the analysis strategy used by Kuru and Wojtanowicz (1988) is very similar although the experimental verification does not seem to have been as thorough as that conducted by Detournay with Glowka's data (1989).

Falconer and Detournay use a similar approach to characterise normal system operation, by aligning points representative of bit performance in the plane of variables. This straight line is the analogue of a rock failure criterion under global load. In the remainder of the document, we will refer only to Falconer's "roller-cone bit" terminology. Warren's mechanical model, expressed in *response* variables R_D and T_D is then written:

$$T_D = c_1 + c_2\sqrt{R_D} \tag{2.1.7}$$

We see immediately the similarity of (2.1.7) with Coulomb's law, provided that in relation (2.1.2) n is given a value of 2, which corresponds to the "normal" regime. However, it is probably more correct to position (2.1.7) with respect to a simple Amonton type interface friction law.

These normal conditions are represented on Figure 2.1.5 by a straight line.

We assume that neither the *rock* parameters determining the characteristics of the straight line (ratio τ/σ of the shear strength to the compressive strength) nor the *bit condition* parameters are affected during the observation interval; only the two *response* components R_D and T_D (or S and E) vary according to the disturbances on the *input*: weight on bit W and speed of rotation N, which are nevertheless already mean values. Variations in rock "hardness" σ may also affect the response.

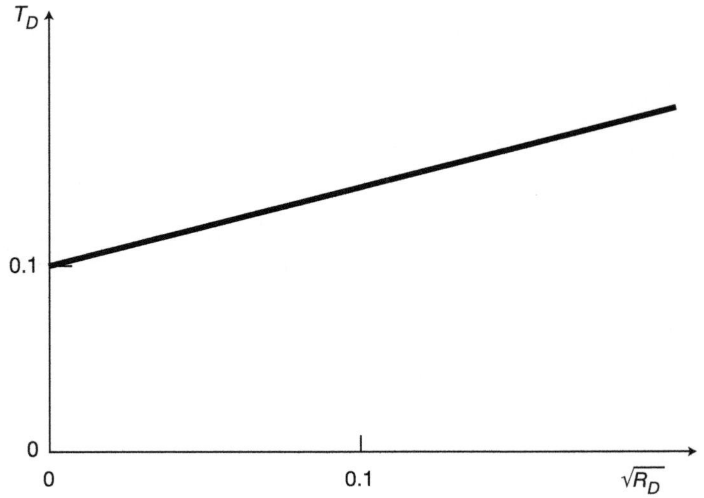

Figure 2.1.5

Normal conditions in the $T_D - R_D$ plane.

The slope representative of normal operating conditions lies between 0.07 and 0.15 for "reference shales". These rocks are considered as favourable to analyse the wear factor, for example, since the ratio τ/σ changes very little within this class, unlike the other rock categories.

The deviations on T_D and on the slope values are attributable to a "non-shale" facies or to transfers within the weight/torque torsor which are generally symptomatic of dysfunctions.

Refer to the publication by Falconer and Defourny for the quantitative effects of the wear on the operating points: the drilling resistance is increased, at constant weight on bit W, according to relation (2.1.3), which decreases R_D. The complementary effect is a substantial drop in T_D indicating a decrease in shearability (Figure 2.1.6).

The criterion indicates the increasing energy contribution with axial thrust, which can be interpreted as increasing "friction" at the borehole wall with confinement.

Another type of improvement was obtained by taking into account bit wear, since this has an important effect on the torque. This correction was therefore introduced in the estimations as an indicator T_D for the torque, sensitive to cutter wear, like the more traditional indicator R_D which corresponds to the axial dimension. Combined use of these two indicators yields a more reliable diagnosis than with a single indicator, especially when studying transitions to degraded or dangerous regimes.

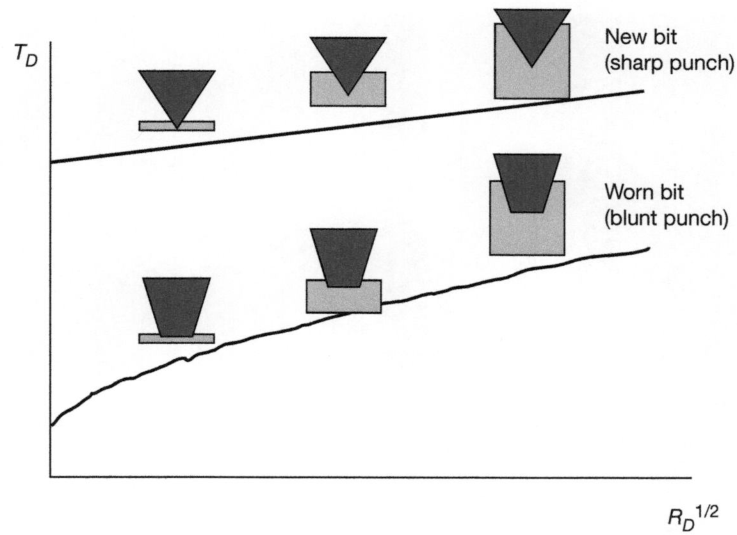

Figure 2.1.6

Effects of wear on the relation of bit behaviour in the plane of reduced variables, especially on the slope and the Y-intercept [Falconer and Defourny, 1992].

2.1.1.3 Cutter wear, a major concern

Studies have only recently turned to an additional parameter, that of cutter *wear*. It represents a determining factor in drilling performance, governing dynamics and rate of penetration. Presence of a wear flat (Figure 2.1.7) cannot be avoided, and the general trend in the past which consisted in exclusively studying new tools is insufficient. Moreover, optimised drilling must take into account bit wear which can be identified. Monobloc bits (PDC type) are much more sensitive to wear. It is quite well-known that new PDC bits are highly sensitive to weight and must treated carefully, especially during the "running-in" phase, to avoid premature damage [Kuru and Wojtanowicz, 1988] [Fear *et al.*, 1995]. This is not the case with roller-cone bits, which are much sturdier in this respect.

Developing PDC bits capable of drilling a highly abrasive rock efficiently, while offering sufficient lifetime, has therefore been a true challenge for bit manufacturers specialised in diamond impregnated cutters, even though this technology is thoroughly controlled. New bits are considered as highly efficient for such conditions but they are so aggressive that they suffer rapid damage if incorrectly controlled when first put into service (gentle control is required during the "running in" phase). The density of the cutters and their arrangement can be designed so that the bit is less vulnerable initially, but this strategy fails as wear progresses and the tool generally becomes ineffective.

For further reading on this subject, refer to [Geoffroy *et al.*, 1997, 1998a, 1998b]. Studies in the field of elastoplastic rock behaviour laws have focused more specifically on the distribution of forces exerted on the cutter, in particular on the wear flat.

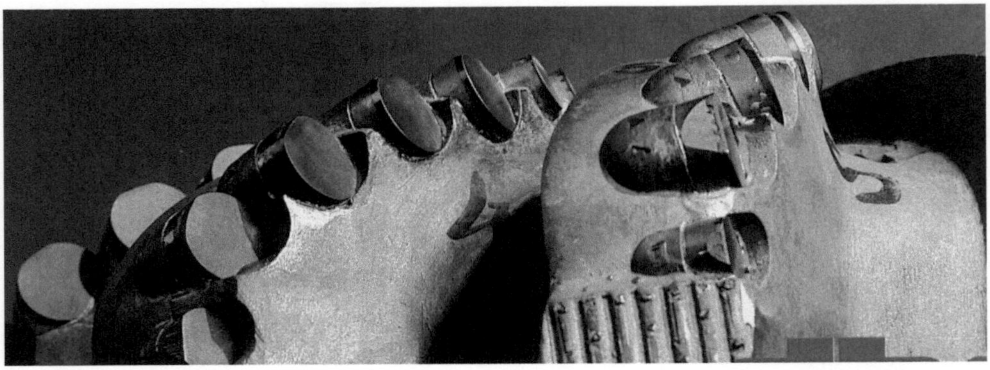

Figure 2.1.7

Example of PDC type bit, showing a wear flat (DBS document).

This action is in line with the studies also conducted by SANDIA [Glowka, 1987, 1989] and the University of Minnesota [Detournay, 1991, 1996]. The distribution of elementary forces on the diamond impregnated cutters and therefore the ability of the bit to cut down the rock have been predicted more accurately. It has led to a strategy for positioning the cutters which is better than that associated with the new bit.

2.1.2 Drillability and detailed analysis of failure mechanisms at the working face

2.1.2.1 Rock cutting mechanisms

Numerous experimental and theoretical studies have been conducted on the fundamental problems associated with the drilling process. They are focused on the mechanical action of a *bit*: in the sense used by Miners, not Drillers! Miners are concerned directly by the cutter which, in fact, they call "a pick" and not by the rock drilling machine, whereas Drillers see the tool more globally and are less concerned by the cutting "studs". This tool punches the surface of a rocky bed, like the tool used in hardness tests. However, the mechanism involved in the hardness test is more one of breaking the rock into numerous fragments of various sizes, rather than a cutting action as such (Figure 2.1.8). The destruction process is therefore more like a rock crushing process. In processes such as these involving a notion of volume of crushed material, the energy – volume relations, resulting in the definition of specific energy, are extremely useful and considered as the significant index of the mechanical efficiency of the drilling process.

In the petroleum industry, "ductile" failure processes are favoured by the high confinement of drilling situations.

Despite the widespread and well-accepted nature of the standard hardness test, the basic understanding of the hardness test in terms of the mechanisms actually involved has only improved slowly. B. Lawn and R. Wilshaw (1975) studied the appearance of the cracks caused by punching with a hard sphere.

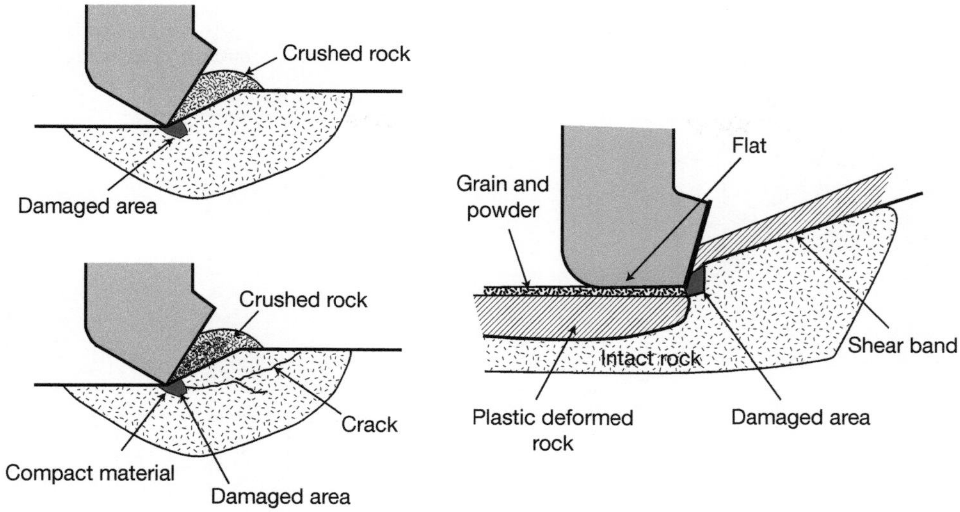

Figure 2.1.8

On the left: Cutting mechanisms observed in sandstone with new PDC blades. On the right: Representation of the damaged region below the flat [Geoffroy, 1996].

Since spherical symmetry (Hertz's theory) is only an approximation which he considers as sufficient, Detournay assumes that there is a core under hydrostatic pressure immediately underneath the punch, extending with the applied load. This core is surrounded by a region in plasticity state obeying a well-defined criterion. Beyond this region we find the elastic matrix. We can demonstrate that this set of states generates a mechanism through which the compressive stress field under the punch is then converted away from the punch into a stress field with a tensile component.

Via a certain degree of plasticity microcracking equivalence, this model introduces the dissipation related to the rock crushing process in the hydrostatic compressive region. Drillers are obviously interested in propagating a crack by simple extension, which generates a well-defined chip. The two dissipation modes present in punching: diffuse mode, associated with plastic deformation of the rock, and localised mode related to propagation of a tensile crack, are both clearly present in Detournay's model (1996).

Various authors have studied the rock cutting mechanism in a rather brittle perspective; for example, Nguyen Minh (1974) for oil applications and Déliac (1986) for mining applications.

2.1.2.2 Shear cutting regimes

Detournay drew up a list of various geometrical cutting situations with associated elementary models [Detournay, 1996].

Detournay considers three mechanisms depending on the penetration per revolution δ (regimes I, II and III below) in increasing order of depth; r designates a characteristic size related to the grain dimension (Figures 2.1.9 and 2.1.10).

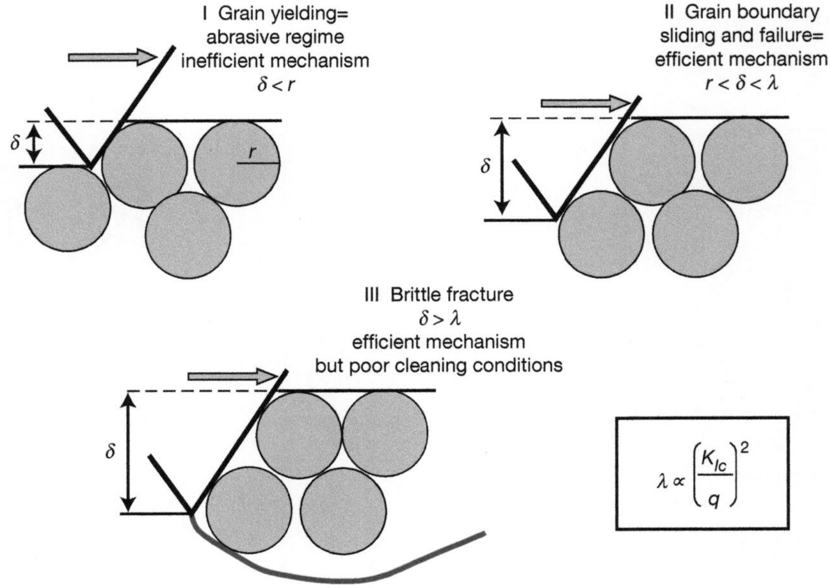

Figure 2.1.9

Cutting mechanisms depending on grain size r and penetration per revolution δ [Detournay, 1996].

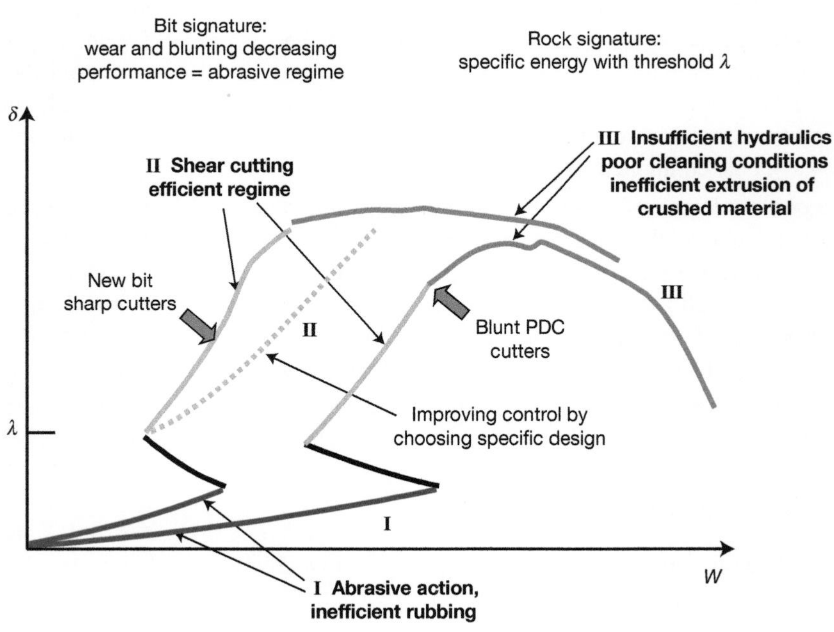

Figure 2.1.10

Bit penetration δ against weight W exerted [Detournay, 1996].

Regime I: $\delta < r$

The first "cutting" regime is that of surface spalling and inefficient abrasion, exhibiting high specific energies. Grain plastification occurs. In the penetration per revolution – weight on bit axis system, this regime marks out a weight efficiency threshold which increases with the size of the wear flat [see also Kuru and Wojtanowicz, 1988]. The order of magnitude of the penetration per revolution is less than that of the grain size.

Regime II: $r < \delta < \lambda$

The second regime corresponds to the cutting mode as such, ductile. This is the preferred mode in practice, which acts efficiently on cohesion, cementation between grains by separation: "grain joint plastification" occurs, consuming less energy than grain plastification.

For the two above mechanisms (regimes I and II), plasticity is not confined.

Regime III: $\delta > \lambda$

The third regime is considered as brittle. It appears from a depth λ proportional to Irwin's length [Detournay, 1996]:

$$\lambda \propto \left(\frac{K_{Ic}}{q} \right)^2$$

where K_{Ic} is the rock toughness (ability of the material to withstand sudden propagation of a crack) and q its unconfined compressive strength. λ is high for ductile rocks. The proportionality factor, mainly related to the bit geometry, not indicated in the previous relation, also governs the ductility. For a sandstone of unconfined compressive strength about 50 MPa, the typical depth λ is about 1 to 2 millimetres.

Regime III can still be considered as efficient, since the hydraulics are sufficient.

The behaviour of the punching model [Johnson, 1985], Figure 2.1.11, follows the well-known progression of "strain stages" when the force applied increases: elasticity, contained plasticity, unconfined flow.

Damjanac and Detournay recently (1995) resumed studies on stage II with contained plasticity (Figure 2.1.12).

This analysis can also be compared with that conducted by Kendall (1978) who, considering "the complexity of compression failure modes" according to the area over which the compressing force is applied (*w* plate width) compared with the cross-sectional area of the test piece (*d*), also identifies three failure mechanisms. Spalling, which prevails at low values of *w/d*, axial splitting with elastic failure criterion involving K_{Ic} for intermediate values, plastic buckling or barrel failure at high values.

2.1.2.3 Consequences and applications

Using the almost straight line relation between force and penetration per revolution in ductile regime II (Figure 2.1.10), the mechanical properties of the rock can be identified from a simple core. This property led to the design of the RSD (Rock Strength Device) whose operation is similar to that of a cutting tool.

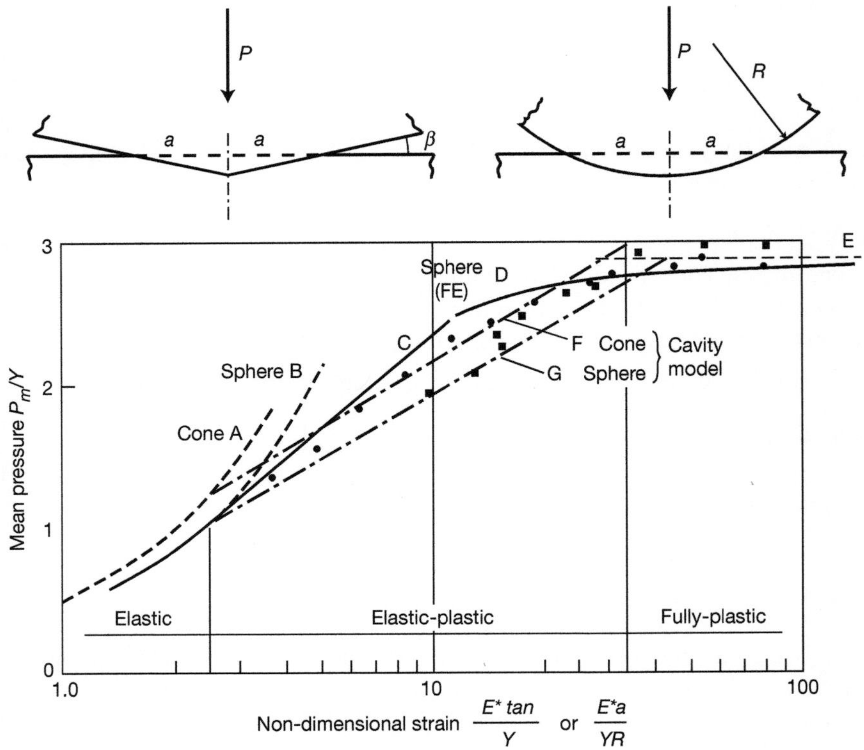

Figure 2.1.11

Johnson model (1985).

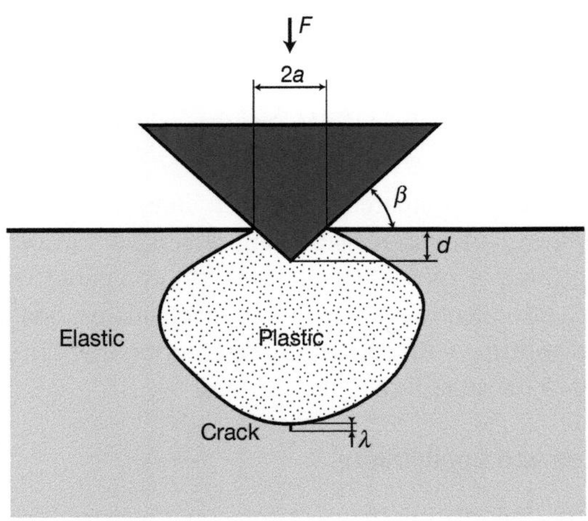

Figure 2.1.12

Normal indentation of a wedge into rock
[Damjanac and Detournay model, 1995].

In practice, the characteristic of the global rock drilling bit system in the required operation is also a straight line, in the plane with the weight on the x-axis and the penetration per revolution on the y-axis. The slope of the line is representative of the intrinsic specific energy of the rock. It is very high for a sandstone, the specific energy being very low. If the weight on bit is not precisely controlled, we rapidly enter abrasion mode (regime I) or mode III which, due to accumulation of chips, soon causes cleaning problems.

In contrast, the specific energy is high in shales, increasing with confining pressure. The bit is controlled better with the weight.

The weight efficiency threshold is related to the presence of a more or less pronounced wear flat. It is worthwhile converting this information in a plane containing the reduced variables weight and torque and interpreting using Detournay's formalism.

The link between the specific energy and the cutter angle of attack may also be considered; if very large, energy is lost while cutting, but there is better tool control.

The cutting test indicated previously can be used to obtain the rock properties, provided that we are in regime II.

2.1.2.4 Punching models for bit behaviour

Determination of a model of the punching process and especially of the plastic zone development phase is critical in order to obtain a good understanding of the tensile crack initiation and propagation stage. Using the cavity expansion model as a basis, Detournay demonstrated (1996) that the process is controlled by a single scalar γ measuring the "degree of elastoplasticity" and governing the size of the plastic zone:

$$\gamma = \frac{2(K_p + 1)G tg\beta}{\pi q}$$

β is the punch angle,
q the unconfined compressive strength,
G the shear modulus,
K_P a coefficient related to the angle of friction.

This factor γ is quite similar to the typical strain ε (load-dependent) described by Johnson (1985) for the spherical indentation (Figure 2.1.11):

$$\varepsilon = \frac{E^* a}{YR}$$

where E^* is the equivalent modulus of elasticity, R the indentation radius, a the contact circle radius, Y the yield stress.

For the conical indentation, this strain, independent of the load, is:

$$\frac{E^* tg\beta}{Y}$$

The punch is conical, with half-angle at the top $\pi/2$-β. The radius of the contact circle is a. Sharp cones (β close to $\pi/2$) favour plasticity, as do small radii R.

E^*/Y is the ratio of the modulus of elasticity to a yield criterion, generally the shear failure modulus. For a highly cohesive rock this ratio will be moderate (several hundred), for a poorly cemented rock it will be higher (several thousand). Note that E^*/Y is the reciprocal of the strain at failure.

The formalism on the transition of mechanisms provides a mechanical explanation of the Johnson or Detournay indices. Including the specific fracture work, it introduces competition between failure and plasticity. This is in fact the meaning of the cutting mechanism boundary λ indicated by Detournay.

2.1.3 Drilling performance including evacuation of cuttings and bit dynamics

Drillability is traditionally studied in terms of steady state regimes, with limited fluctuation of the characteristic quantities of the system.

By definition, the study of dysfunctions concerns transitions, whether due to a poorly managed control parameter – generally the weight, but possibly the speed of rotation or the hydraulics – whose untimely variation triggers the phenomenon, or through the sudden appearance of a rock formation transition for which the current control is unsuitable, thereby inducing an unfavourable effect.

This description is also valid for dysfunctions such as excessive vibration in hard rocks or for an incident related to clogging in soft rocks.

As we have seen in 2.1.1.2, normal behaviour is characterised by an operation line in a suitable plane of variables (Figure 2.1.5) relating mean quantities ($R_D - T_D$) and which suggests a Mohr-Coulomb type failure criterion with, in particular, a steady increase in the normalised torque T_D when the confinement effect R_D increases (Falconer representations for the roller-cone bit, Detournay for the PDC).

A significant departure of the operation points away from the "normal" line may come from:

- (i) A radical modification of the rock behaviour at failure (sudden lithological transition).
- (ii) A vibratory dysfunction during the interaction of the bit with a very hard rock or on the contrary degenerating into clogging/jamming with a soft rock for which the control was not adapted on time. Clogging designates conditions under which the balance of divided solid material (cuttings) is in excess, jamming refers to conditions which are "even more irreversible" for which physicochemistry makes the bonds between aggregates and bit structure even stronger.

(ii) is in fact an exaggerated form of (i).

- (i) In this case, after facies transition, the bit is no longer under optimised conditions allowing efficient drilling but we are nevertheless outside the field of normal operation.
- (ii) In this case, operation is disturbed on leaving the transition and drilling cannot continue in the long term; an inappropriate modification of the control or lack of reactivity by the driller during a facies transition may cause these untimely and often irreversible excursions.

2.1.3.1 Modelling of jamming during drilling

Jamming or clogging is a major problem encountered by drillers [Garnier and van Lingen, 1958] [Hebert and Young, 1972]. The problem occurs when, due to defective hydraulics or unfavourable lithology, rock chips build up at the working face; various descriptive mechanisms have been proposed, some privileging the bit and some the formation as "catalytic" site of the aggregation process. Nevertheless, existing models [Wardlaw, 1971] [Warren, 1984] cannot be used as a predictive base and *a fortiori* as a means of control to avoid it.

One original study consisted in using conservation laws and simplified constitutive relationships [Eronini *et al.*, 1982] to determine the evolution equations giving the global accumulation rate at the working face as a function of time. The critical regimes, as a function of the weight, speed of rotation and flow rate control, were identified; excessive weights quickly lead to clogging, with large cuttings (Figure 2.1.13); insufficient weights result in slow pollution of the working face by fine particles and also induce an inefficient regime that can be qualified as jamming, thereby completing, from a hydraulic point of view, the description of the well-known effect of mechanical efficiency threshold [Putot, 1995, 1997].

Putot's model represents a significant improvement over Warren's semi-empirical representations in case of defective hydraulics since it considers, in a simplified way, the phenomena actually involved. Obviously, the disadvantage is that a large number of parameters must be identified. However they all have a physical significance. The advantage of the analysis concerns a realistic description of the conditions leading to a bit clogging mechanism.

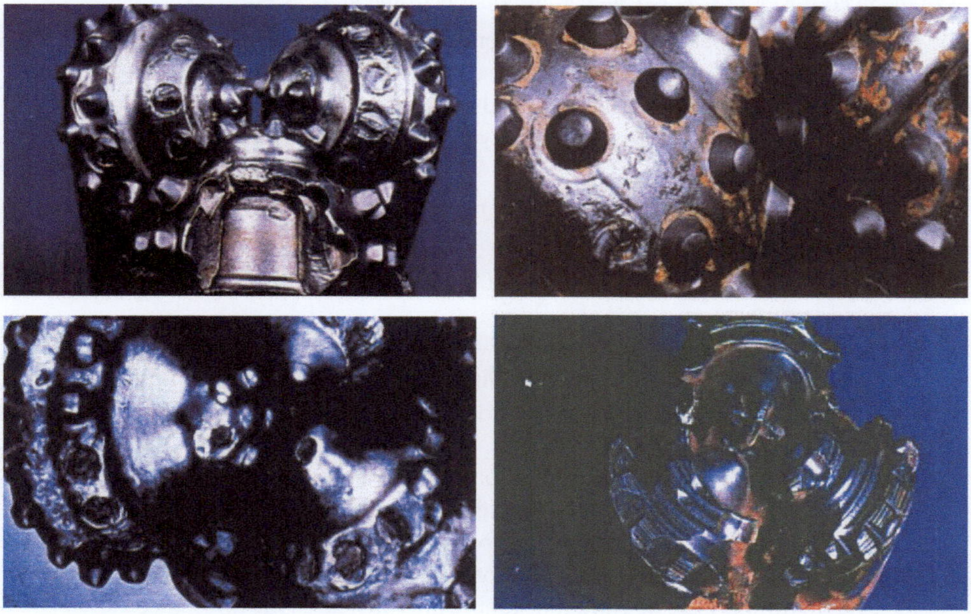

Figure 2.1.13

Examples of damage to roller-cone bits. Top left: broken cone. Top right: worn tooth. Bottom left: broken tooth. Bottom right: jamming.

A dynamic version of the bit coupled with the cuttings concentration has been developed [Putot and Constantinescu, 1997]: the state of the working face has two dynamic variables (in the longitudinal and circumferencial directions) and a concentration variable; the solution of the differential system combining these state variables is as much one of a problem of dynamics as one of a problem of concentration evolution. The influence of the initial conditions and the severity of the ramps forming the loading paths (in particular weight and torque control) have been studied. Compared with the quasi-static version, this model offers highly significant advantages: influence of bit dynamics on concentration evolution – and improved understanding of the drilling system dynamics in case of clogging "problems"; but also, introduction of the rotation degree of freedom and consequently torque taken into account in the response observation criteria. Validation has been carried out semiquantitatively on a GdF Suez site [Putot *et al.*, 2000].

2.1.3.2 Drilling efficiency and vibrations of the drilling system

Symptoms of drillstring vibrations vary widely. Sometimes, the drillstring bounces or rotates so irregularly that the drill team is clearly aware that something is wrong. In other cases, for example in case of transverse vibrations, the anomaly is undetected at the surface and can lead to the failure of the equipment, observed too late!

The problems due to vibrations are varied and often extremely serious. Frequently, the damaged equipment must be brought to the surface, the borehole is irregular and, during coring operations, the samples are damaged (e.g. core discing). Vibrations are correlated with a reduced rate of penetration. Limiting the vibrations generally involves higher costs, due to the need for more sophisticated equipment [Jardine *et al.*, 1994].

It is important to dispel the notion that vibrations may benefit the rock drilling process since, while the rate of penetration may increase in the short term, this comes at the expense of shorter lifetime of the bit and drillstring. Although vibrations such as bit noise are used as a seismic signal while drilling, these are usually of a different frequency and amplitude from the destructive vibrations.

Drillstring vibrations can be divided into three types: axial, torsional and transverse:

- Axial vibrations cause bit bounce and rough drilling, behaviour that destroys bits, damages Bottom Hole Assemblies (BHA), increases total drilling time and may be detected at surface.
- Torsional vibrations cause irregular downhole rotation that fatigues drill collar connections, damages the bit and slows drilling. The vibrations are recognised at the drillfloor by fluctuations in the power needed to maintain a relatively constant rate of surface rotation.
- The more destructive transverse vibrations may be unleashed with no sign at surface. Deep in the hole, the BHA interacts with the borehole wall, generating very violent transverse shocks; this leads to an out-of-gauge and rough hole and the shocks can damage components of the BHA. The bottomhole equipment (optional) used to assess the severity of this type of vibration significantly increases the drilling costs.

Vibrations of all three types may occur simultaneously or one at a time, depending on the case, during rotary drilling and are more violent in vertical or low-angle wells, where the drillstring may move more freely than in high-angle wells with a long offset.

Many studies of drillstring vibrations employ one-dimensional, linear harmonic analysis that largely ignores the dynamics of wellbore wall contact. This approach has proved relatively successful for axial and torsional vibration. Studies have concluded that there are two types of torsional vibrations: transient and stationary. Transient vibrations correlate with variations in drilling conditions such as heterogeneity in the rocks. Stationary vibrations are caused by the natural resonance of the drillstring and are most likely to cause problems.

During rotary drilling, the most recognisable manifestation of stationary torsional vibration is stick-slip. Because of friction between the bit and BHA (Bottom Hole Assembly) and the wellbore and the spring-like nature of the drillstring, the bit actually stops rotating, even though the drillpipe is still being turned at a more or less constant rate on surface. This is known as the stick phase. Sufficient torque is progressively stored up in the system, until it finally overcomes friction and releases the bit which accelerates suddenly, reaching angular speeds far greater than the nominal speed. This is known as the slip phase. This phase may last for several seconds, depending in particular on the length of the drillstring and, to a lesser extent, the mean speed of rotation. The speed then decreases to start another stick phase. These more or less periodic oscillations exert high cyclic stresses on the drillpipe and slow drilling [Pavone and Desplans, 1994].

The high speeds achieved during the slip phase are also one of the causes of transverse vibrations due to centrifugal force [Théron *et al.*, 2001].

Axial vibrations are also excited, since the axial drilling rate depends on the angular speed. Lateral accelerations are frequently 10 times greater than axial accelerations.

However, despite the relation with rotation, analysis of transverse vibrations cannot rely on a harmonic model. This is important since, in the past, it was believed that by "fine-tuning" the drillstring – varying its rate of rotation and weight on bit – to avoid a resonant frequency, transverse vibrations that have been initiated downhole may be stopped. Studies by ANADRILL reveal, on the contrary, that this does not always work.

Field observations have led to the conclusion that transverse vibration problems appear in hard or abrasive formations. Shocks are often seen as the MWD (Measurement While Drilling) or adjacent stabiliser passes from a soft to a hard formation, without changing the speed of rotation.

At the heart of the generation of transverse vibrations is a self-sustaining interaction between the rotating drillstring and the wellbore [Théron *et al.*, 2001]. Once initiated, this interaction is hard to stop. Once shocks are observed downhole, varying the speed of rotation will often affect only the number and severity of the shocks. Shocks are not eliminated until the speed of rotation is reduced to a much lower value than the rate at which they commenced. Thus, significant shocks may be sustained at all but the lowest speeds of rotation.

Variation in transverse vibration with lithology is due to changes in the rock's coefficients of friction and restitution. As friction between the drillstring and rock increases, more energy per impact is transformed from rotary to transverse motion. In some cases, friction may be exaggerated through the digging action of stabilisers clashing with the borehole.

The coefficient of restitution determines what proportion of the kinetic energy at impact is absorbed by the formation. Low restitution results in significant energy absorption. Because limestones and sandstones have high coefficients of friction and restitution, they are more likely to generate high shocks than soft shales – an intuitive result.

2.2 BOREHOLE STABILITY DURING DRILLING

2.2.1 Instabilities and drilling

2.2.1.1 Description of drilling modes

Two drilling modes can be used: OBD (Over Balanced Drilling) and UBD (Under Balanced Drilling). OBD is a procedure used to drill wells where the mud pressure is greater than the pore pressure. Most wells are drilled using this method. In UBD, the mud pressure is less than the pore pressure and production takes place while drilling. In the remainder of this document, we will implicitly assume that OBD is used.

2.2.1.2 Borehole instability mechanisms while drilling

When a well is drilled, the rock surrounding the borehole must take up the load previously supported by the rock that has been removed. This results in the development of a stress concentration at the borehole wall. If the rock is not strong enough, the wall will fail.

In OBD, the mud density is determined to prevent a fluid influx from the formation ($P_{mud} > P_{pore}$). If the mud density is too low compared with the mechanical stability criterion, compressive failures could occur, possibly causing the well to collapse. The mud density must therefore be adjusted according to the most restricting criterion (pore pressure or compressive stability).

The mud density cannot be increased infinitely, however, since if it is too high, tensile failures caused by hydraulic fracturing will develop in the rock (Figure 2.2.1).

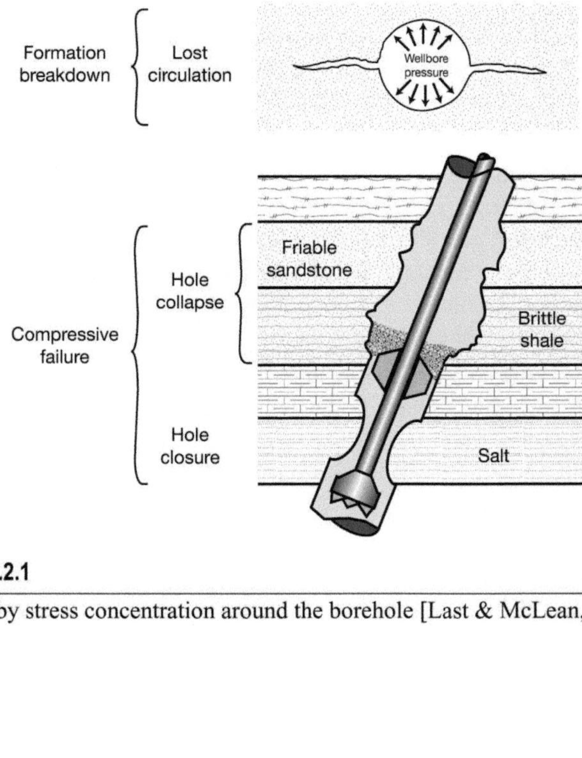

Figure 2.2.1

Failure by stress concentration around the borehole [Last & McLean, 1995].

The mud pressure must therefore lie within a window (Figure 2.2.2) whose:

- lower limit is the larger of two values, which are the pore pressure P_{pore} and the minimum mud pressure necessary to ensure stability of the walls $P_{collapse}$,
- the upper limit is given by the fracture propagation pressure P_{frac} in the terrain, i.e.:

$$max\ (P_{pore}, P_{collapse}) < P_{mud} < P_{frac}$$

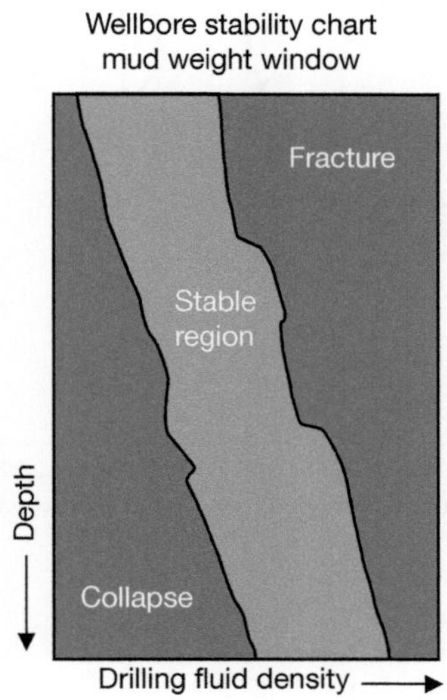

Figure 2.2.2

Mud weight window.

The mud weight window may be very narrow when the – abnormal – pore pressure is close to the hydraulic fracturing pressure, a situation observed in numerous parts of the world, for example the North Sea [Ward *et al.*, 1994] and the Potwar Basin in Pakistan [Kadri, 1991].

The true mud pressure in the borehole depends on the static weight of the mud column increased by the dynamic effect of the flow (known as ECD – Effective Circulating Density), together with occasional fluctuations as the drillstring moves (pistoning or suction). In view of these fluctuations, the borehole stability conditions are often borderline.

In OBD, the aim is to work at a pressure slightly above the lower limit for several reasons:

- to detect impregnated reservoirs,
- to limit the risks of fracturing and mud losses,
- to increase drilling performance.

Compressive failures are highly dependent on the *in situ* stress field. Failures occur progressively. Spalling appears towards the minor stresses in the first phases followed by stabilisation with borehole breakout or continued enlargement of the borehole if the spalls come loose from the wall and fall off (Figure 2.2.3).

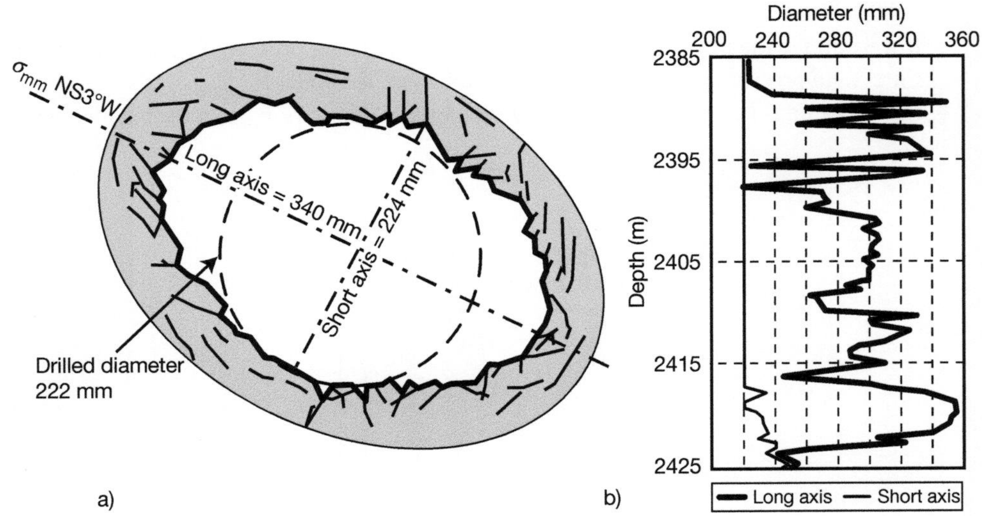

Figure 2.2.3

Schematic illustration of spalling in the damaged area around a borehole. Detection of borehole breakout using a Calliper [McLellan, 1994].

Maury and Sauzay (1987) identify more precisely three destabilisation mechanisms corresponding to three different stress models at the wall (σ_a = axial stress, σ_θ = tangential stress, radial stress = mud pressure):

- mode A, with "intermediate" axial stress ($P_{mud} < \sigma_a < \sigma_\theta$), indicated by the presence of narrow, elongated spalls,
- mode B, with "intermediate" tangential stress σ_θ ($P_{mud} < \sigma_\theta < \sigma_a$), producing curved spalls concentric with drilling,
- mode C, shear failure when the mud pressure is high enough to open the natural fractures ($\sigma_\theta < P_{mud} < \sigma_a$), generating small wedge-shaped spalls (Figure 2.2.4),
- mode D, extension failure.

The temperature variations induced by the circulation of mud may have considerable influence on borehole stability [Maury, 1994]. Local cooling (at well bottom) releases the compressive stresses around the borehole, possibly leading to temporary stability which will disappear if the temperature rises (e.g. if the mud circulation stops). Inversely, local heating (at the top of the well) may cause instabilities. Maury mentions that, in some cases, mud cooling devices have proved an efficient way of reducing instability. It is important to know the temperature distribution, or at least have an estimation (Figure 2.2.5). A number of software programs capable of predicting the temperature variations in the borehole are now available for operators.

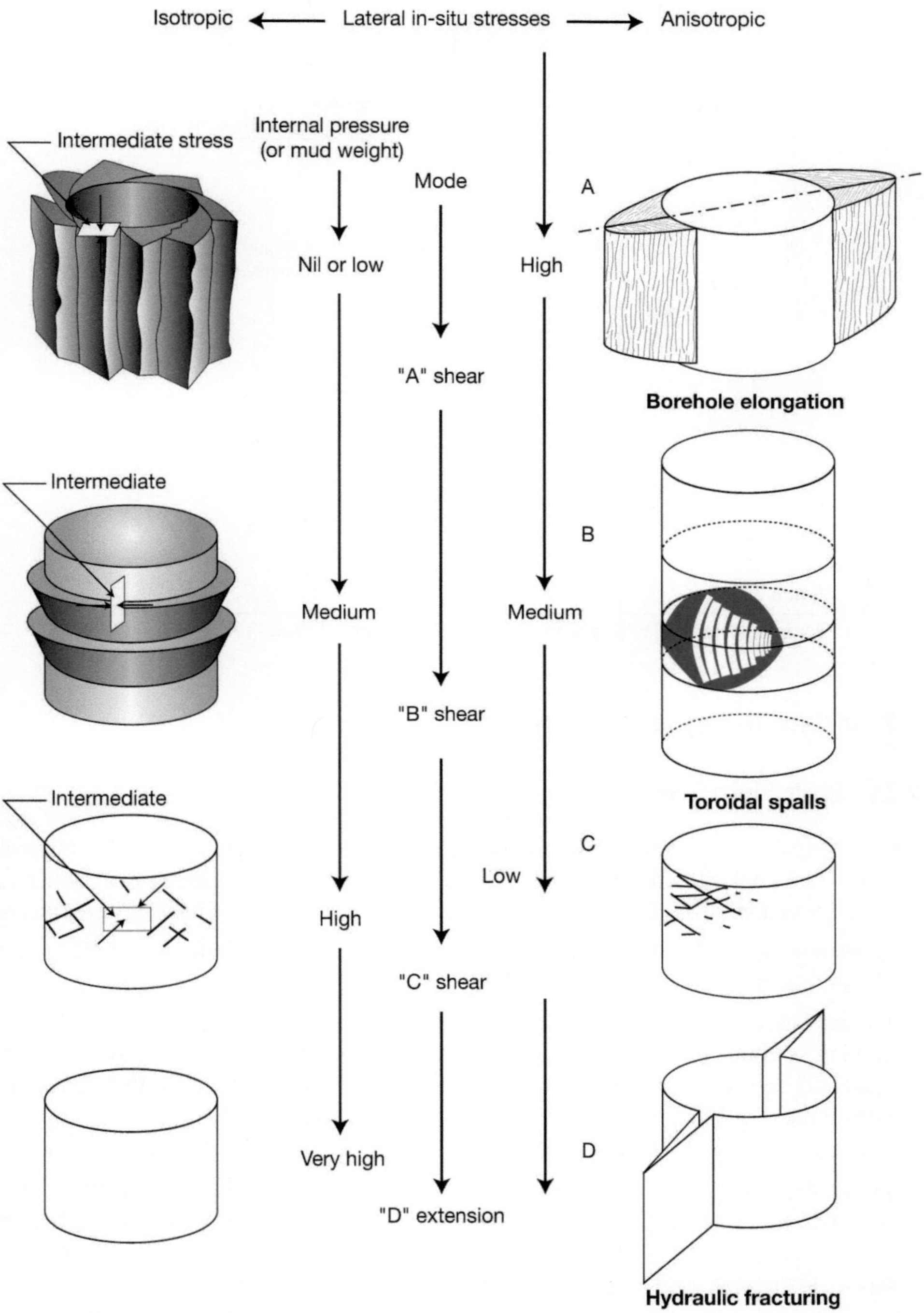

Figure 2.2.4

Failure modes and spalling process [Maury & Sauzay, 1987].

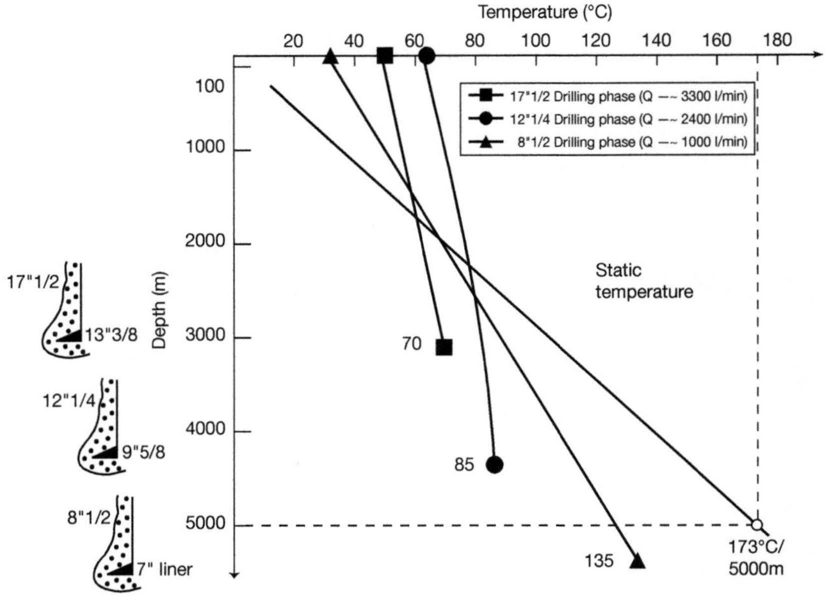

Figure 2.2.5

Temperature profiles during mud circulation [Maury, 1994].

2.2.2 Instabilities by formation type

2.2.2.1 Shale formations

Shales are highly compressible sedimentary rocks, composed of interstratified clay minerals (kaolinite, illite, chlorite, montmorillonite) which can absorb water through their structural network. The cationic exchange capacity is a good estimation of the swelling smectite content.

Shales represent 75% of the drilled formations and are responsible for over 90% of well stability problems [Steiger & Leung, 1992].

If water-based drilling muds are used, migration of water into the shales is likely to induce high swelling stresses. *In situ* stresses, however, limit water penetration and hydration generally remains restricted to the surface [Onaisi *et al.*, 1994]. Note that diffusivity increases if the rock is exposed to a mud hotter than the formation.

In case of imbalance between the formation and mud water compositions, ionic migration occurs as well as water migration. We may therefore consider the propagation of three fronts – pressure, water and ion – which may locally modify the effective stresses and the mechanical strength of shales [Fam and Dusseault, 1998; Chen *et al.*, 2003a]. These fronts propagate slowly and instability phenomena may only appear after several days [Mouton *et al.*, 1998]. They worsen over time, however (Figure 2.2.6). Reduction in local mechanical properties and increase in pore pressure favour failure and produce spalling, enlarging the borehole and creating numerous associated problems.

The choice of drilling fluid is important for the swelling shales; a high electrolyte concentration may reduce crystalline swelling but it is recommended to use polymers and seal narrow fractures with gilsonite.

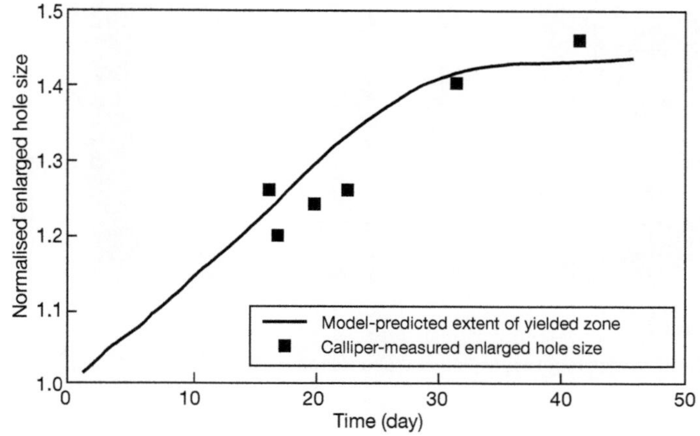

Figure 2.2.6

Hole enlargement against time in a shale formation [Hawkes and McLellan, 1997].

2.2.2.2 Unconsolidated formations (erosion)

These formations are less unstable than shales. Numerous incidents may occur while drilling, however:

- caving caused by hydraulic erosion,
- deposit of a thick cake, risk of differential pressure sticking,
- mud losses due to excess pressure in the annulus,
- splitting, sliding for shales.

Problems are also encountered in arctic regions when drilling in the permafrost. The heating caused by mud circulation considerably reduces the strength of the layers.

2.2.2.3 Fractured/cracked formations

Limestones, dolomites and other brittle formations may be naturally fractured or become fractured while drilling. Invasion of the fracture network may cause loosening of the blocks and their erosion (Figure 2.2.7). Blocks of all sizes may then break away from the walls and fall into the annulus. Damage increases with time and may cause the well to collapse.

These fractured formations may have Unconfined Compressive Strengths (UCS) greater than 50 MPa and the dominant failure mode may be due to friction of the fractures [McLellan and Cormier, 1996, Chen *et al.*, 2003b]. The walls deteriorate over time, as can be seen on acoustic or electric images.

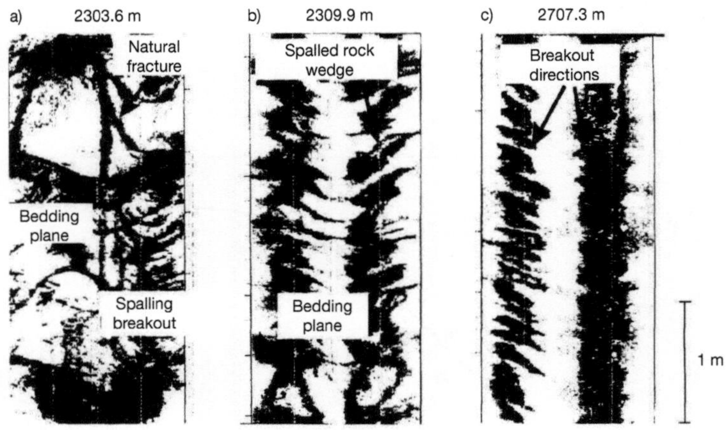

Figure 2.2.7

Borehole ultrasonic images in a cracked material [McLellan and Cormier, 1996].

In this type of formation, changing from a water-base mud to an oil-base mud has no effect and increasing the mud density favours mud invasion of the fracture network [Santarelli *et al.*, 1992]. This problem is solved by adapting the mud rheology.

In some extreme cases, fault invasion will reduce the local effective stresses and the shear stresses may cause lateral displacement of the borehole axis [Maury, 1994] with the consequences we can well imagine: drillstring sticking, or even breaking (Figure 2.2.8).

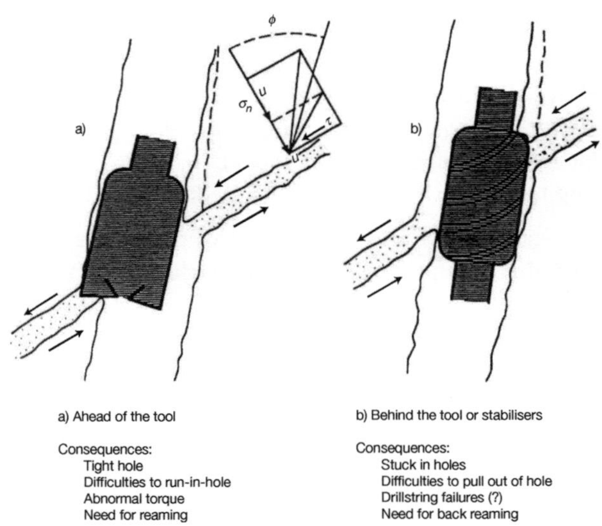

Figure 2.2.8

Lateral displacement of the borehole axis induced by shear relaxation along pre-existing faults or fractures [Maury, 1994].

2.2.2.4 Creeping formations (salt or plastic shales)

The properties of evaporitic rocks (salt) vary depending on the component minerals (halite, carnallite, sylvinite, kieserite and bischofite). Firstly, they are highly water-soluble (bischofite is 10 times more soluble in hot water than halite (NaCl – table salt). Washouts beyond the range of calliper logs of have been observed with water-base mud [Williamson *et al.*, 1997].

The behaviour of salt under stress is similar to that of a highly viscous fluid. The strain rate depends on stress (depth), temperature, mineralogical composition, water content and the presence of impurities such as clays. Rates of 2 cm/hour have been recorded.

Some formations may be highly fractured, resulting in mud losses.

The operational problems have been identified: fluid kicks (brine, oil, gas, salt), losses (filtration, fractures), drillstring sticking, casing collapse (Figure 2.2.9).

Large plastic strains may be observed in soft shales.

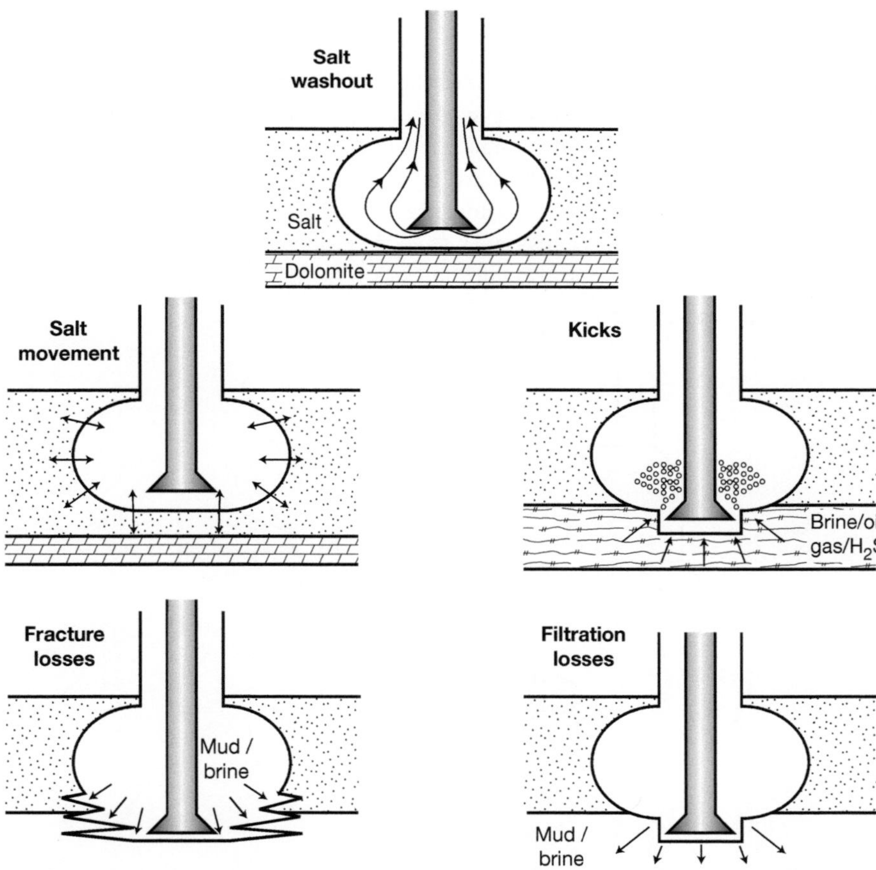

Figure 2.2.9

Instabilities in salt [Williamson *et al.*, 1997].

2.2.2.5 Formations in tectonic zones

Stress distribution in tectonic zones differs from that observed in calm zones. The vertical stress may be less than the horizontal stresses, which may be highly anisotropic. Variability is generally high in the presence of faults.

In case of instability, spalls and breakouts spread in the direction of the lowest stresses, for example parallel to the mountain ranges in the Columbia and Alberta foothills [Last and McLean., 1995, Plumb *et al.*, 1998].

2.2.3 Stability calculation

2.2.3.1 Analytical solution to the elasticity problem

The analytical solution to the elasticity problem for an inclined wellbore in a system exhibiting anisotropic horizontal stress is described in numerous books and publications [Fjaer *et al.*, 1992; Charlez, 1997]. The main equations given by Aadnoy and Chenevert (1987) will be mentioned below.

The following assumptions are made: linear elasticity and isotropy of the *in situ* rock mechanical properties. We assume that the principal stresses in the formation are: σ_v vertical stress, σ_H maximum horizontal stress, and σ_h minimum horizontal stress. Stresses near the wellbore are described in a coordinate system where the z-axis is parallel to the wellbore and the y-axis is horizontal (Figure 2.2.10). Transformation of coordinates (σ_v, σ_H, σ_h) to (x, y, z) is carried out by two rotations, a rotation "a" around the σ_v axis and a rotation "i" around the y-axis. Angle i represents the wellbore deviation and angle a the azimuth with respect to the maximum horizontal stress.

P_w represents the mud pressure in the wellbore of radius R.

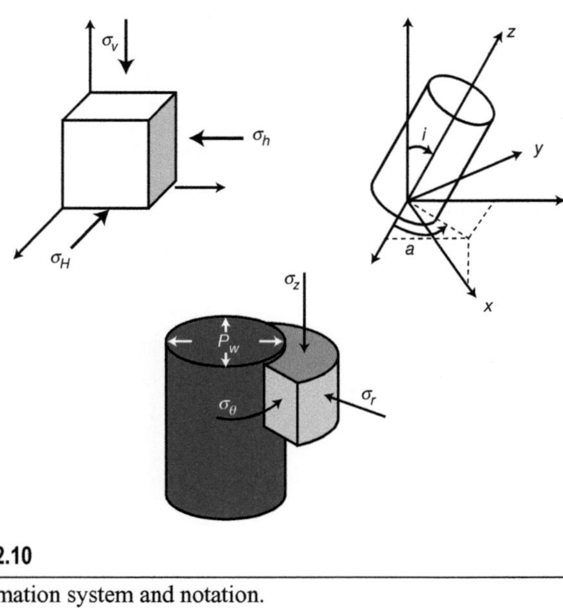

Figure 2.2.10

Transformation system and notation.

The stresses in cylindrical coordinates are given by the following expressions as a function of the distance r to the centre of the wellbore.

$$\sigma_r = \left(\frac{\sigma_x + \sigma_y}{2}\right)\left(1 - \frac{R^2}{r^2}\right) + \left(\frac{\sigma_x - \sigma_y}{2}\right)\left(1 + 3\frac{R^4}{r^4} - 4\frac{R^2}{r^2}\right)\cos 2\theta$$

$$+ \tau_{xy}\left(1 + 3\frac{R^4}{r^4} - 4\frac{R^2}{r^2}\right)\sin 2\theta + \frac{R^2}{r^2} p_w$$

$$\sigma_\theta = \left(\frac{\sigma_x + \sigma_y}{2}\right)\left(1 + \frac{R^2}{r^2}\right) - \left(\frac{\sigma_x - \sigma_y}{2}\right)\left(1 + 3\frac{R^4}{r^4}\right)\cos 2\theta - \tau_{xy}\left(1 + 3\frac{R^4}{r^4}\right)\sin 2\theta - \frac{R^2}{r^2} p_w$$

$$\sigma_z = \sigma_{zz} - 2\nu\left(\sigma_x - \sigma_y\right)\frac{R^2}{r^2}\cos 2\theta - 4\nu\tau_{xy}\frac{R^2}{r^2}\sin 2\theta$$

$$\tau_{r\theta} = \left[\left(\frac{\sigma_x - \sigma_y}{2}\right)\sin 2\theta + \tau_{xy}\cos 2\theta\right]\left(1 - 3\frac{R^4}{r^4} + 2\frac{R^2}{r^2}\right)$$

$$\tau_{rz} = \left[\tau_{xz}\cos\theta + \tau_{yz}\sin\theta\right]\left(1 - \frac{R^2}{r^2}\right)$$

$$\tau_{\theta z} = \left[-\tau_{xz}\sin\theta + \tau_{yz}\cos\theta\right]\left(1 + \frac{R^2}{r^2}\right)$$

With

$$\sigma_x = \left(\sigma_H \cos^2 a + \sigma_h \sin^2 a\right)\cos^2 i + \sigma_v \sin^2 i$$

$$\sigma_y = \left(\sigma_H \sin^2 a + \sigma_h \cos^2 a\right)$$

$$\sigma_{zz} = \left(\sigma_H \cos^2 a + \sigma_h \sin^2 a\right)\sin^2 i + \sigma_v \cos^2 i$$

$$\tau_{yz} = 0.5\left(\sigma_H - \sigma_h\right)\sin\left(2a\right)\sin i$$

$$\tau_{xz} = 0.5\left(\sigma_H \cos^2 a + \sigma_h \sin^2 a - \sigma_v\right)\sin 2i$$

$$\tau_{xy} = 0.5\left(\sigma_H - \sigma_h\right)\sin\left(2a\right)\cos i$$

At wellbore wall r = R, so the equations become:

$$\sigma_r = p_w$$

$$\sigma_\theta = \sigma_x + \sigma_y - 2(\sigma_x - \sigma_y)\cos 2\theta - 4\tau_{xy}\sin 2\theta \; - p_w$$

$$\sigma_z = \sigma_{zz} - 2\nu\left(\sigma_x - \sigma_y\right)\cos 2\theta - 4\nu\tau_{xy}\sin 2\theta$$

$$\tau_{r\theta} = 0$$

$$\tau_{rz} = 0$$

$$\tau_{\theta z} = 2\left[-\tau_{xz}\sin\theta + \tau_{yz}\cos\theta\right]$$

In an anisotropic horizontal stress field ($\sigma_{H\,max} > \sigma_{h\,min}$), concentrations of maximum circumferential stress σ_θ are observed on the wellbore walls in the direction of the minimum horizontal stress (Figure 2.2.11). We can therefore see that breakouts will occur in these zones, depending on the hardness of the rock.

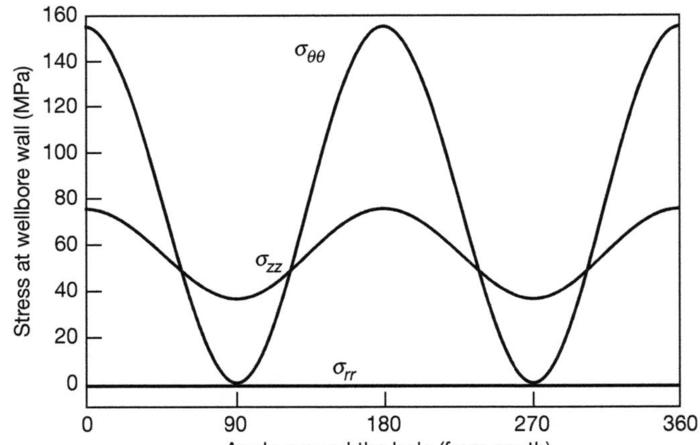

Figure 2.2.11

Stresses σ_θ around a vertical wellbore under an east-west compressive stress field [Zoback *et al.*, 2003].

σ_z and σ_θ are not principal stresses, which are calculated as follows:

$$\sigma_1 = \frac{\sigma_\theta + \sigma_z}{2} + \frac{1}{2}\sqrt{(\sigma_\theta - \sigma_z)^2 + 4\tau_{\theta zz}^2}$$

$$\sigma_2 = \frac{\sigma_\theta + \sigma_z}{2} - \frac{1}{2}\sqrt{(\sigma_\theta - \sigma_z)^2 + 4\tau_{\theta zz}^2}$$

$$\sigma_3 = p_w$$

Any compressive failure criterion can be applied to estimate the collapse failure: Mohr Coulomb, which does not take into account the intermediate stress, Von Mises, Drucker Prager, etc. [McLean and Addis, 1990].

2.2.3.2 Wall instability model

Various software programs are now available to calculate wall stability for any wellbore angle and any stress field.

These tools run rapidly on PC with a simple poroelastic model, sometimes taking into account fluid diffusion against time as well as thermal and chemical effects.

They are often conservative [Van Oort *et al.*, 2001]. Rocks are not purely elastic and continue to have a load bearing capacity after failure. Operators therefore use these linear elastic models for fast analysis and optimisation of the wellbore trajectory, but still require more sophisticated models involving finite element codes, with poroelastoplasticity models (Cam-Clay or Laderock type).

These models have demonstrated in particular that horizontal wellbores are mechanically more stable if they are drilled parallel to the maximum horizontal stress [Zoback *et al.*, 2003].

2.3 SAND PRODUCTION

Sand production designates transport of grains by the production fluid, from the reservoir rock into a wellbore.

In gas and light oil reservoirs, or in offshore production, above a certain proportion, sand production causes a number of undesirable problems such as damage to wellbore pumps and wellhead erosion (a leak may be created by sandblasting in less than an hour in a gas well), plugging of perforations or even total invasion of the production column. Several methods are available to address these problems, but they are expensive to implement and often reduce the production rate.

In heavy oil reservoirs, if sand production prevention systems are used, the oil flow rates are considerably reduced (from 7/15 m^3 a day to 0.5 m^3 a day, for example) and the well becomes uneconomical. In this case, it is acceptable to produce oil and sand simultaneously, as with the Cold Heavy Oil Production with Sand (CHOPS) method which may prove profitable under certain conditions. Due to the high viscosity and low production rates, pump and valve erosion problems are minor and the sand can be transported up to the surface without difficulty using progressive cavity pumps. If thermal stimulation is used, for example Steam Assisted Gravity Drainage (SAGD), the oil viscosity is lowered and systems must be installed to prevent sand production.

In all cases, therefore, anticipating the onset of sand production is a major challenge.

The physical problem is extremely complex and involves numerous parameters: rock and reservoir characteristics, wellbore completion type and production conditions. Some triggering mechanisms are well-known, for example damage to soft formations by drilling or depletion, and damage under the effect of hydrodynamic forces; other, such as water cut, erosion at the surface of cavities, etc., are much less well-known.

2.3.1 Types of sand production

We could speak of "solid" production since this term is less restricting than "sand" which refers to a geological and grain size classification. According to Dusseault and Santarelli (1989), chalk and coal may also produce solids, ranging from sands to silts and clays. The fact remains that, in the large majority of cases, it is sand production which occurs.

The term sand production includes a wide range of phenomena. On the basis of field observations, Veeken *et al.* (1991) suggested three types of solid production: transient, continuous and catastrophic.

2.3.1.1 Transient solid production

Transient solid production refers to a solid concentration declining with time under constant well production conditions. This phenomenon is frequently observed during clean-up after perforating or acidising, after pullout or after water breakthrough. The rate of sand production may vary considerably, for example a sand volume of 1 to 200 L over a period of 1 to 500 hours.

2.3.1.2 Continuous production

In a large number of fields, continuous levels of sand production are observed. What constitutes an acceptable level of sand production depends on operational constraints (erosion, separator capacity, sand disposal, lift, etc.). For example, the following levels can be considered as acceptable:

- 6-600 g/m^3 in oil wells,
- 16 kg/Mm^3 in gas wells.

The well may nevertheless fill up with sand and require regular cleaning. Sand production increases the volume of perforations and may even create large cavities around the casing.

2.3.1.3 Catastrophic solid production

Catastrophic solid production refers to events where a high influx of solids, sometimes several dozen m^3, causes the well to choke and/or die.

The following correlations are observed:

- The grain size distribution of the sand produced is identical to that of the formation.
- Sand production is more likely to occur with poorly consolidated, low compressive strength ($Rc < 2$ MPa) rocks at shallow depth.
- Sand production is triggered or increases with reservoir depletion, when the effective stresses inside the reservoir increase.
- Sand production occurs when the pressure drawdown (difference between pore pressure far from the wellbore and pressure in the flowing well) reaches or exceeds a certain critical threshold; the critical pressure drawdown is associated with the rock mechanical strength. Moricca *et al.* (1994) speak of total pressure drawdown, the difference between the initial pore pressure far from the wellbore and the pressure in the flowing well. Field observations show that sand production continues while the stress conditions which triggered it remain and stops when these conditions disappear.
- Watercut is often correlated with sand production.

2.3.2 Mechanisms involved

During production, stress changes around the well and the more or less constant flow of gas or oil create instabilities which crumble the rock forming the reservoir and the fluid flow brings the material into the well.

2.3.2.1 Initiation of sand production

At microscopic scale, sand production is initiated when grains of sand are broken off the walls of the perforations under the effect of the production fluid. Particles can only be dislodged if the force applied to the sand particle by the fluid is greater than the sum of the shear strengths available at the point of contact with the adjacent particles.

The type of cement between the grains plays an important role. A silty or argillaceous cement has a low shear strength; a carbonate cement may exhibit strengths that are very high but extremely strain-sensitive; cohesion losses may occur at very low strains.

Solid production is often associated with watercut. One of the explanations put forward is the decrease in mechanical strength through loss of capillary effects, even though the cohesion generated by these forces is very low [Vaziri *et al.*, 2002]. For a silty-argillaceous cement, watercuts could also weaken the cementation bonds.

At macroscopic scale, the zone eroded probably extends radially and solid production may result in modifications to the production and the reservoir (possible creation of cavities, wormholes, connection by facture between wells, subsidence). Progression of the zone eroded is generally unknown, however.

2.3.2.2 Special case of CHOPS

Cold production of viscous oils (> 5,000 cP) is generally uneconomic due to the low potential of the well associated with the viscosity of the fluid to be produced. In some viscous oil reservoirs, however, a number of operators in Canada have observed that the productivity of wells run with acceptable or even encouraged sand production (CHOPS) was much greater than the theoretical productivity calculated assuming radial flow around the wells. In addition, final recovery of the oil from these reservoirs is also abnormally high. Two mechanisms have been identified to explain this phenomenon, one hydrodynamic, the other geomechanical. The first is related to the fact that heavy oils tend to foam, while the second is explicitly related to sand production [Renard *et al.*, 2000].

Sand production has been made possible by the use of progressive cavity pumps. This type of pump can in fact be used to produce effluents containing a high proportion of solids. Sand production is high at the beginning, 30% to 50% of the total production. It then decreases progressively, stabilising at a few per cent after a period of several months. Total sand production reaches hundreds or even thousands of cubic metres after several years. Figure 2.3.1 illustrates the oil and sand productions recorded on the Frog Lake field.

Poorly consolidated heavy oil reservoirs in Canada show a clear correlation between sand production and higher productivity: sand production leads to the formation of areas of increased permeability around the wells. Sand production mechanisms are still poorly understood, however, and the geometry of these high-permeability areas is undetermined.

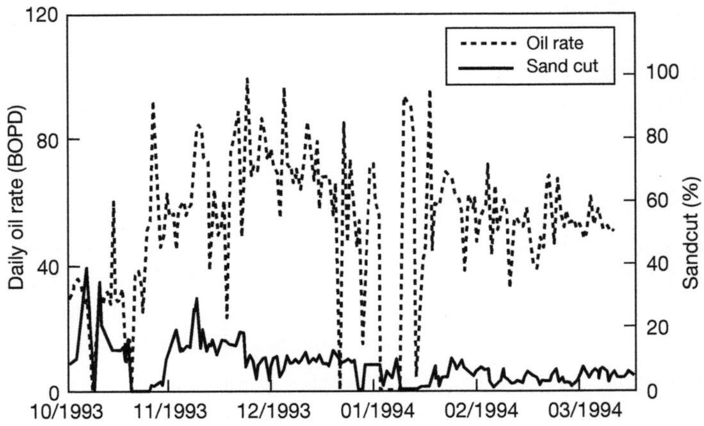

Figure 2.3.1

Oil and sand production on the Frog Lake field [Huang *et al.*, 1998].

In case of continuous sand production, the "decompacted" area around the well grows to provide the sand. There are two possible scenarios (Figure 2.3.2). The first assumes that the eroded zones around the perforations increase spherically and merge to form a liquefied cylindrical zone around the well. This is known as the compact growth scenario.

In the second scenario, each zone or at least some of them start to extend longitudinally in the producing layer, creating a network of highly permeable channels (wormholes). In this case, it is impossible to accurately define the boundary between the intact zone and the remoulded zone of the reservoir.

The physical mechanisms are the same in both cases, however: sand erosion and liquefaction are induced by the hydraulic gradient.

Some authors even suggest combining the two models [Dusseault *et al.*, 1998], [Tremblay *et al.*, 1998]. Well-to-well tracer tests confirmed that there were highly permeable channels between wells [Elkins *et al.*, 1972]. Some authors believe that these channels are wormholes, others that they are fractures.

Laboratory experiments demonstrate that both erosion models are possible [Tremblay *et al.*, 1998 Yalamas *et al.*, 2004, Servant *et al.*, 2007].

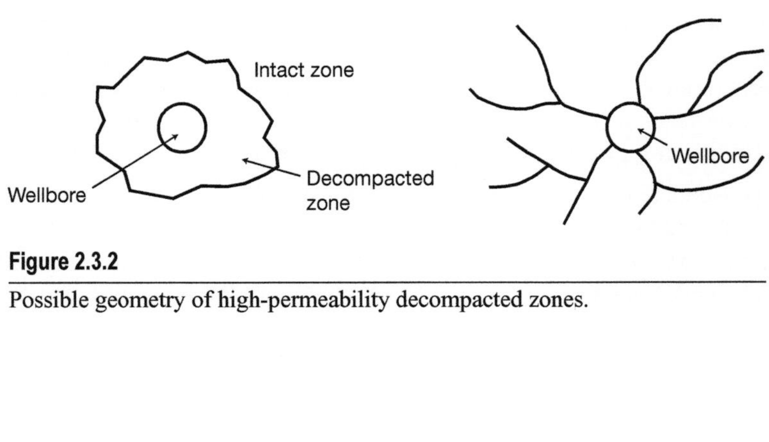

Figure 2.3.2

Possible geometry of high-permeability decompacted zones.

2.3.2.2.1 Assumption of compact growth in the remoulded zone

Geilikman and colleagues assume a double simultaneous flow of the viscous fluid and the granular matrix. The two obviously interact [Geilikman and Dusseault, 1994, 1997].

In their model, the fluid is assumed to be incompressible and the skeleton viscoplastic. The rheological properties of the skeleton, in particular its viscosity, are assumed to vary with porosity. The model is assumed to have radial symmetry.

The transition zone between the intact zone and the remoulded zone is certainly complex and the cylindrical approximation is only used to simplify calculations (Figure 2.3.3).

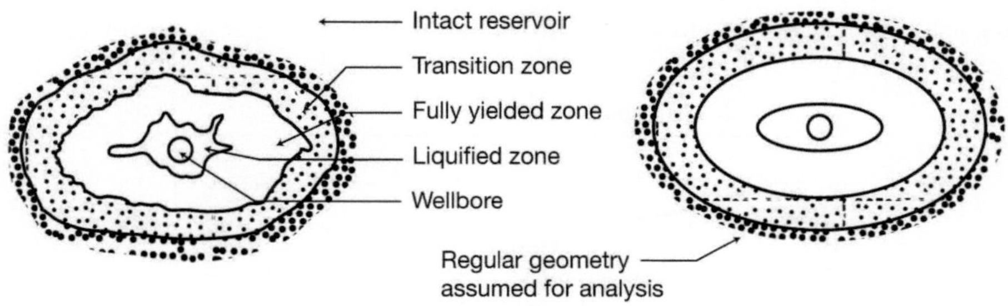

Figure 2.3.3

Growth of a deconsolidated zone [Dusseault *et al.*, 1998].

Geilikman suggests that if the fluid pressure in the well drops below a critical threshold, the shear yield condition becomes fulfilled and a decompacted, yielded and flowing zone starts to grow around the well. The pressure drawdown (difference between reservoir pressure and wellbore pressure) is the key parameter.

Solid production is assimilated to a phenomenon of unravelling of fabric (similar to unravelling of knitting). The boundary between the intact zone and the decompacted zone moves against the direction of flow.

The front propagates as long as the pressure drawdown remains above a certain threshold. Since the reservoir pressure decreases with time, pressure drawdown also decreases until it eventually drops below the critical threshold. At this stage, the front stops propagating.

Application of the model proposed yields productivity gains (about 2), less than those currently found in the literature (10 to 100).

For a 10 m thick reservoir which produced 500 m^3 of sand, the eroded zone would have a diameter of 5 m over an erosion area of 300 m^2. The remoulded zone is probably much larger, however.

2.3.2.2.2 Assumption of growth in wormholes

This approach assumes that the structure of wormholes is stable, roughly cylindrical and of cross-section relatively constant over their length. Wormholes are filled with a slurry flowing slowly and their tips continue to propagate. This approach is supported by numerous experimental studies conducted at the ARC [Tremblay *et al.*, 1995, 1996, 1997, 1998] and by the analogy with the internal erosion models observed in earth dams. Wormhole diameters vary considerably (from 5 cm to 1 m).

Dusseault identifies two cases:

(1) the number of wormholes is constant, it does not increase with the distance to the well,
(2) the wormholes split and create a 3D network, where the density of wormholes remains constant.

In the first case, the equivalent permeability decreases with the distance to the well. In the second case, the permeability remains high up to a boundary zone which corresponds to the wormhole displacement front. From the production point of view, it will be impossible to distinguish between compact growth and growth by a network of wormholes (Figure 2.3.4).

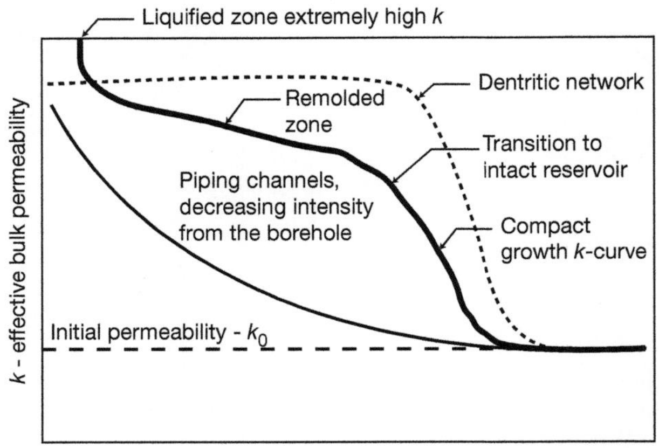

Figure 2.3.4

Effective permeability in different cases [Dusseault *et al.*, 1998].

The contact area for the wormholes could be much larger than for compact growth. Using the same case of 500 m^3 of sand produced, if the wormholes have a diameter of 20 cm, the contact area would be 800 m^2 and the wormholes 16 km long. The smaller the wormhole diameter the greater the contact area, but the larger the diameter the closer we move towards the compact growth model.

Small wormholes containing a viscous slurry will cause head losses along the tube. This limits the pressure gradient that can act at the wormhole tip and, as a consequence, the length of the channel. It is currently impossible to quantify these effects, since there are no methods available to calculate the number of wormholes and therefore estimate the slurry displacement velocity and the pressure drop. It seems unlikely that long wormholes exist, which would suggest a network of dentrites.

According to Dusseault, during the first cold production phase, the remoulded zone is small and the sand production rate high. The radius of curvature of the zone is small and the local pressure gradients are high. Local perturbations are rapidly attenuated.

During the next phase, the remoulded zone is large and any geometric perturbation of the surface zone may lead to the development of a stable wormhole (Figure 2.3.5). This is pure speculation, however. One argument in favour of late "wormholing" comes from the fact that links between wells have been observed when one of them is mature.

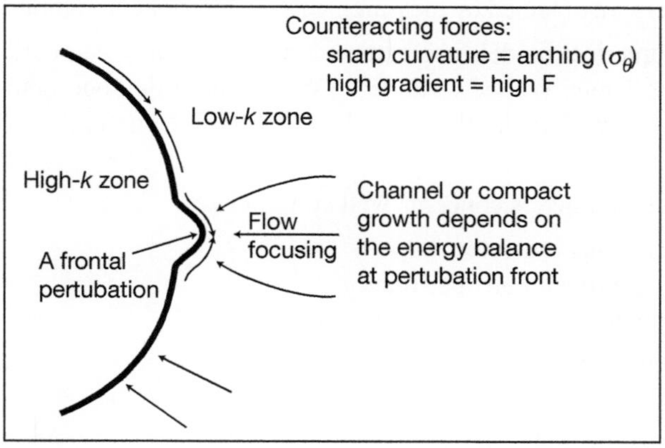

Figure 2.3.5

Frontal perturbation of the decompacted zone [Dusseault *et al.*, 1998].

2.3.3 Sand production prediction

2.3.3.1 Parameters affecting solid production

Numerous parameters can affect solid production. They cannot all be used in the prediction methods since some are difficult to record or understand.

A first series of parameters concerns the reservoir characteristics:

- reservoir thickness,
- porosity,
- type and composition of the fluid (gas, oil, water),
- petrophysical characteristics
  - k rock intrinsic permeability,
  - k_{ro} and k_{rw} relative permeabilities to oil and water,
  - μ_o, μ_w oil and water viscosities,
  - Sw water saturation,
- *in situ* stress field
  - local or default stress tensor,
  - σ_v *in situ* vertical total stress determined by default by the depth and the density of the overlying rocks,
  - σ_h and σ_H *in situ* horizontal stresses,
  - Pr pore pressure at a large distance from the well (decreases during depletion).

A second series of parameters concerns the mechanical characteristics of the reservoir rock:
- *Rc* unconfined compressive strength,
- *C* cohesion,
- φ internal angle of friction.

These parameters are obtained by measurements on cores or correlations with log analyses.

Some recent numerical models use parameters that can only be obtained by conducting special triaxial tests, Thick Walled Cylinder (TWC) tests carried out up to failure under isotropic or triaxial stress with radial fluid flow [Cerasi, 2005, Nouri *et al.*, 2004, 2006, Papamichos *et al.*, 2008].

A third series of parameters concerns well completion:
- well orientation and diameter,
- completion type (open hole, perforations),
- perforation characteristics (orientation, density),
- *Rp* perforation radius,
- *h* perforation length.

2.3.3.2 Empirical methods from in situ measurements

These methods rely only on critical values of parameters above and below which a prevention system is no longer required. They are only valid locally, for a given reservoir:
- critical depth (no problem if $z > 2\,100$ or 3 600 m for example) [Veeken *et al.*, 1991],
- critical value of the compression wave transit time (no sand production if $\Delta tc > 90$ to 120 μs/ft [Veeken *et al.*, 1991], 80 μs/ft [de Sa *et al.*, 1989],
- critical value of the ratio *G/Cb* [Tixier et *al.*, 1975] (*G*: dynamic shear modulus, *Cb*: bulk compressibility) from the sonic and density logs,
  if $G/Cb > 0.8\ 10^{12}$ psi^2, there is a low probability of sand production,
  if $G/Cb < 0.7\ 10^{12}$ psi^2, there is a very high probability of sand production.

The critical depth criterion is the most imprecise since, at a given depth, the rock characteristics may differ from one well to another.

These methods are not suitable for predicting an evolution over time of the reservoir fluid physico-chemical properties (change of water/oil ratio) or calculating the effect of overload of the reservoir grains.

Sand flow test

- The sand flow test increases the rate of flow and therefore the pressure drawdown up to sand production. This method is based on the observation that solid production continues while the stress conditions which triggered it remain and stops when these conditions disappear.
- The quantity of sand produced is measured in a sand trap arranged on the production line (sieve in a container) or, better still, by a piezoelectric or sonic sensor.
- This prediction method is only suitable with homogeneous reservoirs.

2.3.3.3 Rock failure around a wellbore

Depending on the evolution of the stress state around the wellbore, the granular rock disaggregation mechanism is associated with compressive (shear) or tensile failure.

Compression models are required when the tangential stress around the wellbore becomes greater than the rock shear strength. This condition can be triggered by pressure drawdown or general depletion of the reservoir.

Tensile failure occurs when the tensile radial stresses, induced by fluid flow, exceed the rock tensile strength.

For an analytical approach, a model of spherical or cylindrical shells around the cavity formed by perforation can be adopted (Figure 2.3.6).

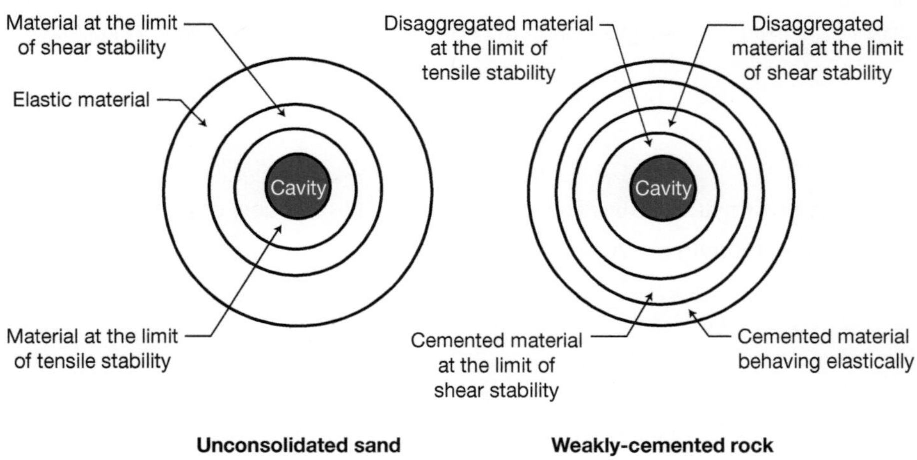

Figure 2.3.6

Model for an analytical approach [Perkins and Weingarten, 1988].

Starting from the centre of the perforation, we observe successively:

- the cavity,
- a shell of disaggregated material at the limit of tensile stability,
- a shell at the limit of shear stability (with possibly one part disaggregated and the other not, depending on the material cementation),
- the elastic virgin material.

The various mechanisms coexist in fact: compressive failure occurs in depleted reservoirs, tensile failure occurs around the cavities (perforations) when the drag forces on the sand particles exceed the cohesive forces with the adjacent grains.

With this type of approach, two critical values can generally be identified:

- a critical reservoir pressure resulting in local shear failures,
- a critical pressure drawdown resulting in tensile failure.

Global (or shear) failure criterion

Several types of compressive failure model have been published:

- elastic-brittle model,
- elastoplastic model,
- poroelastoplastic model, with positive strain hardening, etc.

These models, developed for wellbore stability studies, are highly sensitive to the choice of plasticity criterion and failure criterion.

The models must be calibrated on *in situ* and laboratory tests. A critical reservoir pressure is generally obtained.

Tensile failure criterion

There is also a large number of models. The tensile stability criterion can be expressed in terms of normalised pressure gradient at the borehole wall [Veeken *et al.*, 1991]:

$$H = \frac{\partial P}{\partial\left(\frac{r}{Rp}\right)}$$

where r is the investigation radius and Rp the radius of the borehole or perforation.

The stress equilibrium is written:

$$\frac{\partial \sigma_r}{\partial r} = \frac{\sigma_\theta - \sigma_r}{r} \quad \text{for a cylindrical cavity,}$$

and

$$\frac{\partial \sigma_r}{\partial r} = \frac{2(\sigma_\theta - \sigma_r)}{r} \quad \text{for a spherical cavity,}$$

where σ_r and σ_θ are the total radial and tangential stresses.

We obtain $\sigma_r = \sigma_r' + P$ et $\sigma_\theta = \sigma_\theta' + P$

For a cylindrical cavity, we obtain $\frac{\partial \sigma_r}{\partial r} = \frac{\partial \sigma'_r}{\partial r} + \frac{\partial P}{\partial r} = \frac{\sigma'_\theta - \sigma'_r}{r}$

For $r = Rp$, $\sigma' r = 0$ and therefore $\frac{\partial \sigma'_r}{\partial\left(\frac{r}{Rp}\right)} + \frac{\partial P}{\partial\left(\frac{r}{Rp}\right)} = \sigma'\theta$

Thus

$$\frac{\partial \sigma'_r}{\partial\left(\frac{r}{Rp}\right)} = \sigma'\theta - H$$

A tensile radial stress develops if $\frac{\partial \sigma'_r}{\partial\left(\frac{r}{Rp}\right)}$ is negative, i.e. if $H \geq \sigma'_\theta\,(r = Rp)$

For a spherical cavity, we would obtain $H \geq 2\sigma'_\theta\,(r = Rp)$

Since σ'_r $(r = Rp) = 0$, the maximum σ'_θ is limited, by approximating, by the unconfined compressive strength Rc.

We obtain the limiting values:

$$H = Rc \text{ for a cylinder,}$$

$$H = 2Rc \text{ for a sphere.}$$

Assuming uniform permeability conditions and a stable rate of flow, we obtain a critical pressure drawdown value for initialisation of sand production.

$$Pdd = H \log(Re/Rp) = Rc \log(Re/Rp) \text{ for a cylinder,}$$

$$Pdd = H = 2\, Rc \text{ for a sphere.}$$

Other expressions for the critical pressure drawdown have been proposed by Bratli and Risnes (1981), Perkins and Weingarten (1988) and Ramos *et al.* (1994).

The formulae proposed are very difficult to validate on *in situ* measurements, since only the total flow rate and the global sand production are known *in situ*. We must assume isotropic perforations and a homogeneous formation, which is far from being the case.

It is therefore not surprising that the results of *in situ* measurements do not always correlate with these approaches.

Veeken shows that a critical pressure drawdown Pdd $= Rc/2$ would be a conservative value, valid only if $Rc > 2$ MPa.

Erosion

The hydromechanical aspects have not yet been sufficiently investigated. The first results demonstrate that the erosion surface is accompanied by significant porosity and permeability changes near the surface [Vardoulakis, 1996].

2.3.3.4 Numerical modelling

Finite element codes are considered to be more precise than analytical tools. Combined with laboratory tests, they improve further knowledge of the instability mechanisms triggering sand production and, in particular, they attempt to explain all phenomena observed. Obviously, a few arbitrary levels remain depending on the type of behaviour law employed and the failure criterion chosen (plastic strain, maximum stress, Mohr Coulomb, Drucker Prager, etc.).

The results are nevertheless extremely sensitive to the types of behaviour law and the assumed failure mechanisms. Different results may therefore be obtained with different models, despite being based on the same triaxial tests. It is extremely important to compare modelling results and *in situ* data [Vaziri, 2004].

Simulations with sophisticated material models taking into account the post-peak behaviour of the rock, similar to a soil, supply high-quality predictions for materials of compressive strength between 1 MPa and 10 MPa. They require good knowledge of the rock material with controlled strain triaxial tests conducted at large strains. If triaxial tests have not been conducted, the calibration parameters can be obtained from a database, using as input the unconfined compressive strength and some petrophysical data.

Shao and Marchina (2002) developed an approach based on the macroscopic damage theory. The zone around the wellbore which contains heterogeneities (wormholes, cavities, etc.) is replaced by a homogeneous "damaged" zone (Figure 2.3.7). The change in porosity is associated with an increase in permeability. At the same time, the mechanical properties progressively deteriorate. Numerical simulations with finite element codes demonstrate that the evolution curves obtained are the same as those observed *in situ* for the sand and oil productions.

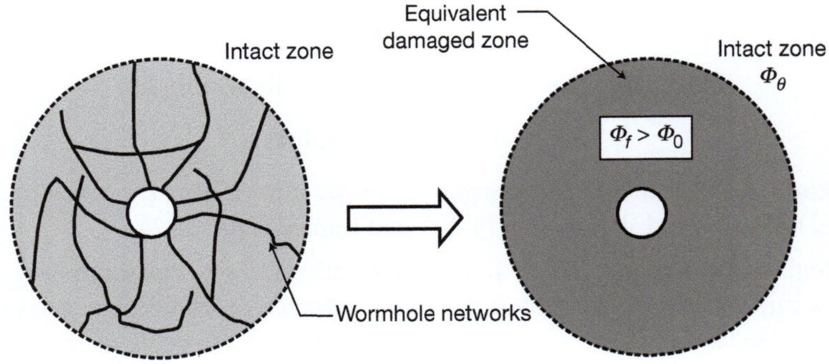

Figure 2.3.7

Illustration of wormholes and equivalent damaged zone [Shao and Marchina, 2002].

Yalamas *et al.* (2004) then Servant *et al.* (2007) developed an original poromechanical approach, without initially prejudging the erosion model. They identify two zones, an intact reservoir zone with poroelastic behaviour and a sand-oil slurry zone, behaving as a Poiseuille fluid whose viscosity varies with the proportion of sand transported. No assumptions are made *a priori* on the geometry of the boundary between these two zones. An algorithm is used to predict the evolution of the boundary as a function of the combined poromechanical and hydraulic conditions (Figure 2.3.8).

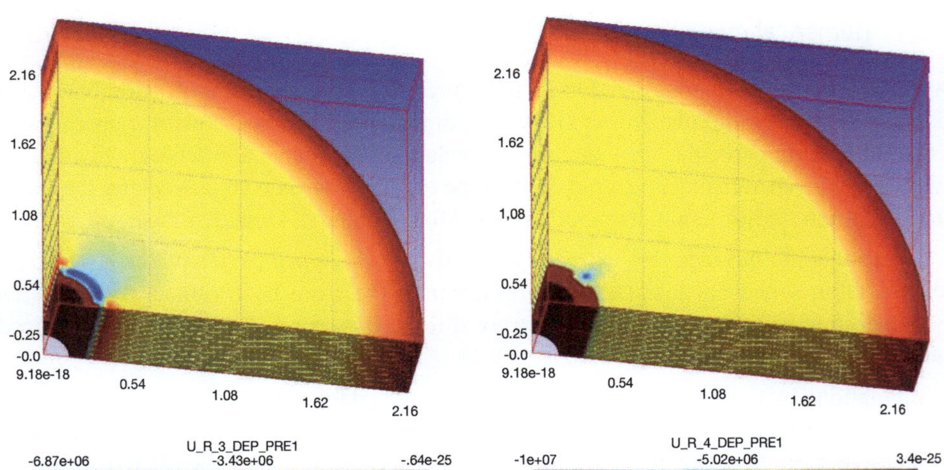

Figure 2.3.8

Eroded zone around a well [Servant *et al.*, 2007].

2.3.3.5 Experimental validation of sand production prediction

Two types of experiment can be carried out, depending on the material concerned:

- sand with no cohesion,
- rock of compressive strength greater than 2 MPa.

Validation is difficult for unconsolidated or poorly consolidated sand. Some authors study the mechanism opposing the start of sand movement under the effect of fluid flow: the formation of stable arches (arching effect). In this case the experimental cells are oedometric type with a hole (or holes) at the bottom or a central wellbore.

These experiments demonstrate that the arch formed by the inner surface of the cavity is stable up to a certain characteristic fluid flow rate. When this flow rate is reached the walls of the cavity collapse, leaving behind a new arch and a larger cavity, which can in turn support a higher flow rate before collapsing. The result of a series of failures such as this is either generalised collapse or creation of a stable cavity [Yalamas *et al.*, 2004].

With indurated rock, the tests consist in studying the stability of a cylindrical cavity created in a more or less large sample subject to triaxial stresses plus, possibly, a fluid flow. The simplest is the hollow cylinder test known as the Thick Walled Cylinder (TWC) test, being carried out up to failure under isotropic or triaxial stress. Incipient failure of a perforation may in fact be related to incipient failure of a hollow cylinder. These tests help to calibrate analytical or numerical models.

However, Veeken and colleagues suggested using the Thick Walled Cylinder (TWC) tests with no fluid flow as independent method. They postulate that the maximum vertical effective stress $\sigma'_{v,\,w}$ supported by a horizontal perforation is equal to the failure initiation pressure of a thick walled cylinder, under isotropic external load $\sigma_{twc,\,i}$ if σ_{twc} is the peak failure pressure.

They therefore assume $\sigma'_{v,\,w} = \sigma_{twc,\,i}$.

The effective vertical stress near a wellbore is $\sigma'_{v,\,w} = \sigma'_v + P_{dd}$ where σ'_v = effective stress far from the wellbore and P_{dd} = drawdown pressure.

Numerous tests have demonstrated that $\sigma_{twc},\ i = 0.86{*}\sigma_{twc}$ (the initiation pressure is less than the peak failure pressure).

Consequently $\sigma'_{v,\,w} = 0.86{*}\sigma_{twc}$.

The critical drawdown pressure is therefore $P_{dd} = 0.86{*}\sigma_{twc} - \sigma'_v$.

The various stress regimes (isotropic in laboratory, anisotropic in situ) and the inner/outer diameter ratio seem to have little influence [Veeken *et al.*, 1991].

2.3.4 Sand production prevention

Sand production can be prevented by:

- reducing the effective stresses,
- mechanically supporting the sand,
- consolidating the sand.

2.3.4.1 Reducing the effective stresses

Since sand production is due to failure of the formation under stresses greater than the strength of the material, the first possible solution is to reduce the stresses (mechanical effective stresses and hydrodynamic stresses) down to a value below this strength.

It is impossible to act on the geostatic stresses. The effective stresses can be prevented from increasing by reducing the pressure difference between the reservoir and the bottom-hole, which reduces the flow, therefore the flow velocities and therefore the hydrodynamic stresses.

This method can only be used on materials strong enough to withstand the geostatic stresses caused by presence of the wellbore. In these types of material, sand production only occurs above a certain production rate. It would seem very difficult to apply this method on materials of very low or zero cohesion for which the wellbore strength is already jeopardised and can only be secured through the use of extremely precise drilling mud programmes. In these types of material, sand production occurs at very low oil production rates.

While reducing the flow rate to limit the hydrodynamic stresses means that it is impossible to use the full production potential of the well, it offers the advantage of not requiring heavy investments for well completion. It may also be necessary to reduce the flow rate in order to improve the recovery rate. This method is therefore easy to implement. However, its choice depends on a rigorous financial calculation, especially in terms of the discount rate.

The producing formation of a reservoir is generally far from being homogeneous, especially as regards the strength of its constituent materials. Sand production can be avoided by only allowing the reservoir to produce through its strongest layers. Due to porosity-strength and porosity-permeability relations, the strongest formations are the least permeable. The decision to perforate only the strongest formations will therefore be based on the resulting loss of productivity compared with the cost of the system necessary to prevent sand production if the more permeable zones are perforated.

During production, the effective stresses in the formation increase as the reservoir is depleted due to lower pore pressure in the reservoir with, as a corollary, lower extraction pressure. This phenomenon results in sand production, which only occurs after a certain time in the life of the reservoir. In this case, pressure maintenance techniques can be used to delay sand production. Generally, pressure maintenance is driven by recovery considerations, the fact that it delays sand production is simply a bonus.

2.3.4.2 Immobilising the material

Sand control consists in inserting a filter between the formation and the wellbore, to prevent the displacement of solid material while allowing fluid to flow. Behind this relatively simple principle however, lie a certain number of difficulties.

- The filter must be as porous as possible and as thin as possible to minimise the head losses suffered by the fluid as it goes through. However, the filter must be able to withstand the static and dynamic stresses due to the flow, which implies that it must be made from a non-negligible quantity of solid; a compromise must therefore be reached in view of the operating conditions.

- The minimum size of the particles trapped will therefore depend on the filter pore dimensions (diameter and tortuosity). Due to manufacturing and head loss considerations, filters cannot be produced with pore sizes below a minimum value; hoping to trap all the solids transported by the flow would be unrealistic. All particles smaller than the minimum pore diameter will pass through and will contribute to "sand" production; once again, a compromise based on the operating conditions must be reached.
- Apart from the wear conditions due to production, the highest stresses on the filter are suffered during its installation in the well. These stresses generally define the dimensional characteristics of the filter, especially porosity, and therefore its ability to trap particles of a given size; a compromise must once again be reached.

Despite the fact that the principle of mechanical control sand is extremely simple, detailed analysis of the actual operating conditions is necessary to reach a compromise between theory and practice.

2.3.4.3 Filters used

The filters used are mainly screens installed in the open hole or the casing, the annular space between the screen and the rock (or the casing) being either filled (gravel pack) or not (natural sand pack) by a filtering material (Figure 2.3.9).

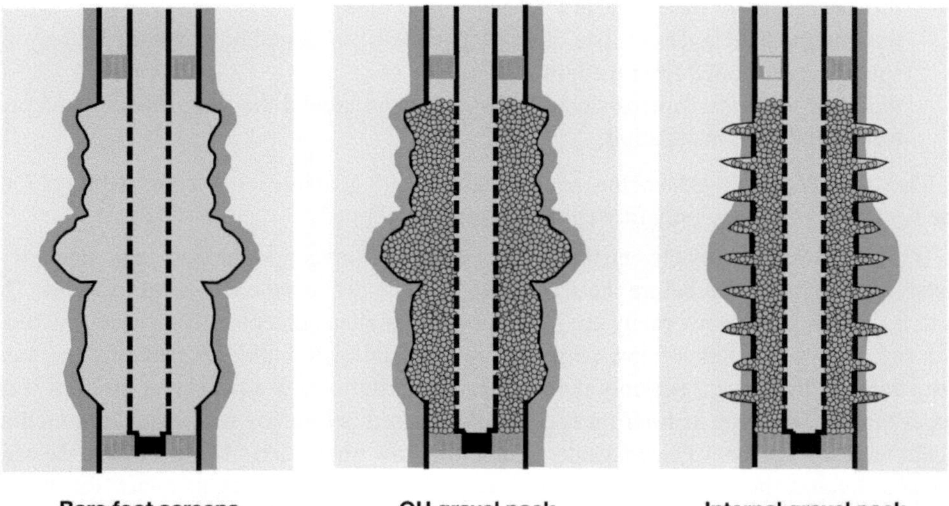

Figure 2.3.9

Sand prevention [J. Alfenore, 2001].

The screens used in open holes can be divided into 5 types:

- perforated or slotted liners,
- wire-wrapped sand screens,
- pre-pack sand screens,
- multilayer sand screens,
- expandable sand screens.

Perforated/slotted liners were the first system used to prevent sand production. These production tubes with API dimensions have slots (or holes) measuring less than 0.4 mm. The area of the holes represents just 3% to 4% of the tube area. These screens have a significant risk of erosion at high flow rates.

Wire wrapped screens are made by tightly winding metal wires around a perforated production tube of API dimensions. Porosity ranges from 6% to 12%.

Pre-pack screens consist of a relatively thin tubular pack, filled with resin-coated gravel (or glass or ceramic beads, metal particles, etc) between two perforated tubes or a perforated API tube inside and a wire-wrapped tube outside. The gravel pack forms the filter medium as such. The filter porosity is about 30%.

Multilayer sand screens are technically the most advanced. They consist of a wire-wrapped and perforated steel tube, one or more layers of filter medium generally made from a woven or non-woven metal material, a wire-wrapped or woven draining layer and an API perforated tube. Porosity of the filter medium can be as high as 70%, thereby reducing the head losses as the fluid goes through.

Expandable Sand Screens (ESS) can expand considerably when a cone-shaped mandrel is forced through them (Figure 2.3.10). They consist of a three-layer screen:

- the inner layer, or base pipe, is a special slotted tube,
- the intermediate layer consists of overlapping strips (wound on the slotted tube) of a specially-developed filter medium,
- the outer layer is a "micro-slotted" shroud wound around the filter medium and protecting it during installation.

They considerably reduce the annular space and guarantee the hole stability. One disadvantage is that they only fit well if the hole is calibrated.

If there are large gaps between the formation and the screen, production deployment will destabilise the formation before the screen can stop the displacement of sand particles. The particles which move most easily are obviously the smallest, however. Consequently, there is a risk that a annulus of low-permeability material will build up between the sand formation and the screen, possibly reducing productivity to such a point that the well is no longer viable. This type of filter must generally be used before any major sand production, except in case of layers of clean sands, with a very low proportion of fines, where the sand displaced against the screen will itself form an additional filter adding its own effect to that of the screen.

The principle of the gravel pack is to create a sand or gravel filter between the producing layer and a screen (generally wire-wrapped) located at the centre of the wellbore. For the filter pack to be efficient, its grain size distribution and that of the sand in the formation must

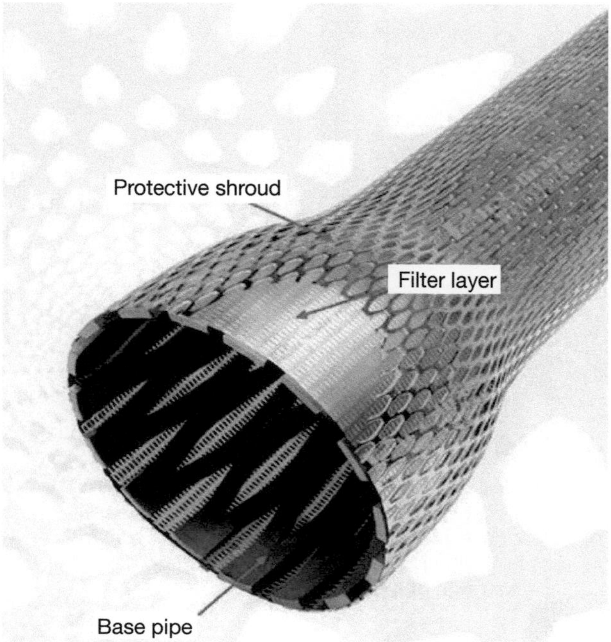

Figure 2.3.10

Expandable sand screen (courtesy Weatherford).

be in a given ratio. The sand or gravel pack is injected into the wellbore with a carrier fluid, which is brought back to the surface once the pack is in position. The installation techniques vary depending on the well architecture: open or perforated cased hole, vertical or horizontal hole. The size distribution of the gravel pack is coarser than that of the reservoir (D_{50} (gravel-pack) $\geq$ 6 D_{50} (formation)).

For optimum efficiency, the sand pack injected must be sufficiently thick; gravel packs are therefore highly suitable for cavities already created at the zones of highest production by earlier sand production or cavities created by perforation.

2.3.4.4 Frac-pack

The principle is to shift the highest hydrodynamic stresses beyond the zone near the wellbore where the highest mechanical stresses occur; this consists in separating the two types of stress so that they affect different zones. To do that, hydraulic fracturing (with tip screen out type) is performed. The tip of the fracture extends beyond the wellbore wall and stops growing at the desired length (Figure 2.3.11). Most of the production occurs via the fracture. Proppants and gravel packs have a number of features in common: injection is similar, their grain size distributions are determined in the same way, both act as filters to prevent sand production.

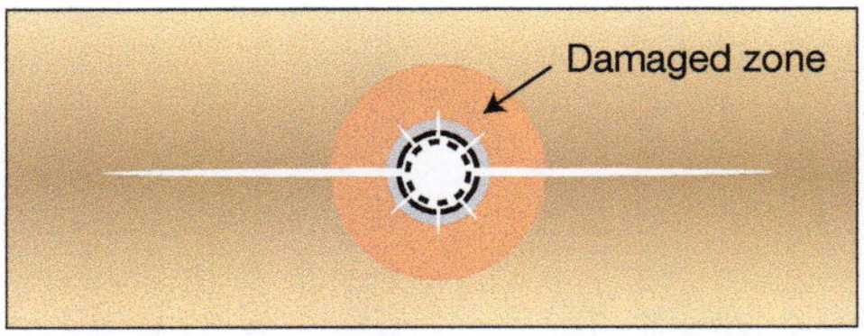

a) Pad pumped without proppant extends initial fracture post near-wellbore damage.

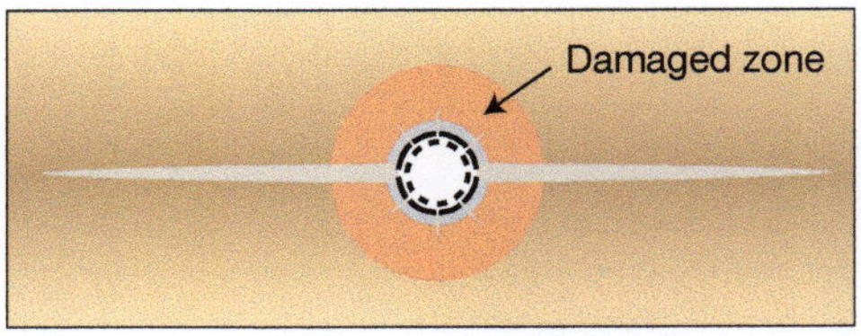

b) Proppant screens out at tip of fracture, arresting fracture extension.

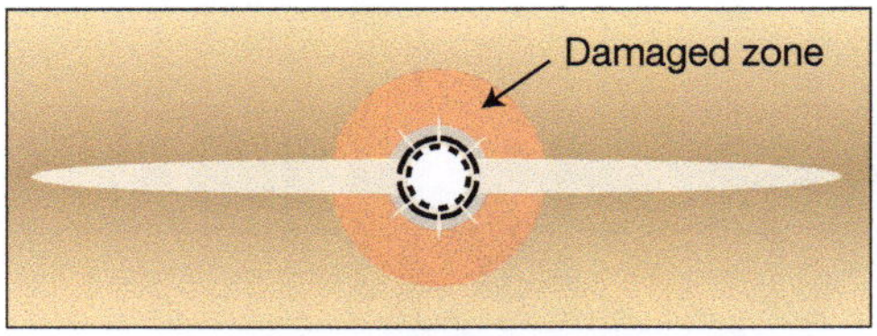

c) Continued pumping with high concentrations of proppant expands fracture width and increases proppant loading. Perforation tunnels, screen/casing annulus, and near wellbore are tightly packed with proppant.

Figure 2.3.11

Frac and pack: create the fracture, inject the proppant, pack [Ellis, 1998].

This method, which can by-pass a damaged zone, is used to replace simple gravel packs which give lower productivity indices.

2.3.4.5 Increasing the strength of the material

Sand production can be prevented by increasing the strength of the material, so that it can withstand the stresses near the wellbore.

The shear strength of granular materials depends on two characteristics: cohesion and friction.

Since it is impossible to modify the nature, shape, etc. of the grains composing the formation, it will be impossible to vary the frictional term to increase the strength of the formation. What can be done, however, is to change the strength of the formation by varying the cohesive term. It would in fact appear that this term is modified, but in the wrong direction, during sand production further to watercut.

To increase the cohesive term, a "binder" is inserted between the grains to give cohesion to the material forming the producing layer, at least over a sufficient thickness around the wellbore (about 0.5 m to 1 m), so that its strength is greater than the stresses induced by the production. Various "chemical" methods have been tested, all involving additional cementation between the grains using a wide range of products: cements, resins, metal deposits, induced crystallisations, etc.

Injecting an "adhesive" between the grains, which should be positioned at the contacts between grains, reduces the porous volume available for the flow of effluents produced, thereby reducing the permeability of the formation near the wellbore, in other words that which has the greatest impact on the well productivity index. While the aim is to minimise the reduction in porosity (permeability), the strength obtained obviously depends on the quantity of adhesive injected. A compromise must therefore be reached between the strength required and the acceptable drop in productivity.

The decision to use a method to increase the strength of the material, despite the resulting drop in well production potential, may be financially justifiable by the fact that well completion is simpler compared with other methods. This method of increasing the strength of the formation material is suitable, for example, with small-diameter wells, where sand screens or gravel packs which require space in the well cannot be used. In contrast, implementation of this method over large areas is complicated.

The well must be prepared before any sand consolidation treatment. In particular, it must be washed and, if possible, all damage due to the production and/or completion operations must be eliminated, to avoid consolidating the damage and making it permanent. The formation is often prepared by prewashing to expel the fluids (water and/or oil) present around the wellbore (zone to be injected), which are incompatible with the treatment. Treatment is then carried out either by single-phase injection: resin plus hardener (catalyst) or by two-phase injection: resin then, after injecting a spacer to clean the excess resin from the porosity, the catalyst. With two-phase treatment (external catalyst), it is important to ensure that the catalyst can reach the resin left in place after injecting the spacer.

2.3.4.6 Sand management

Sand management consists in allowing the well to produce under conditions such that limited sand production is tolerated while avoiding massive sand production. Sand production in surges reduces the skin and cleans the wellbore. Avoiding the use of screens and gravel packs reduces completion costs and increases oil production. However, much research is still required on the conditions of application of this method, to accurately predict the sand flow rates and volumes.

2.4 WELL STIMULATION BY HYDRAULIC FRACTURING

Reservoir stimulation by hydraulic fracturing has become increasingly important since its introduction in the petroleum industry in 1947. The process is used to stimulate reservoirs with poor or low permeability or restore some highly clogged wells and is suitable for a wide range of reservoirs (sandstone, limestone) at depths of up to 8,000 metres.

Currently, about one and a half million oil, gas and water injection wells throughout the world are stimulated by hydraulic fracturing.

Hydraulic fracturing is sometimes associated with other recovery methods:

- acid fracturing of carbonate formations,
- fracturing followed by *in situ* combustion in oil sands or oil shales.

In addition to its possible variants as reservoir stimulation process, hydraulic fracturing has been tested or is used in numerous other industrial applications:

- development of paired wells to allow fluid to flow between them [Boutéca *et al.*, 1983],
- fractures in salt layers, so that water can flow between two wells in order to recover the salt and/or create a storage cavity,
- geothermal technologies,
- coal gasification, methane desorption,
- *in situ* combustion of oil shales,
- storage of industrial pollutants in impermeable layers (shales),
- reinjection of cuttings or sand,
- treatment of polluted soils in the surface layers.

A number of books (or chapters) are dedicated to well stimulation by hydraulic fracturing [e.g. Gidley *et al.*, eds 1989, Charlez, 1997, Economides *et al.*, 1998]. It is impossible to cover all aspects and we will simply provide an overview of the geomechanical aspects.

2.4.1 Description of the hydraulic fracturing process

The objective is to create a permeable drain extending each side of the wellbore by fracturing the rock. A section of the wellbore is isolated with packers. Hydraulic fracturing consists in injecting, at a suitable rate, a more or less viscous fluid at the producing formation until a fracture is initiated from the wall of the wellbore (Figure 2.4.1).

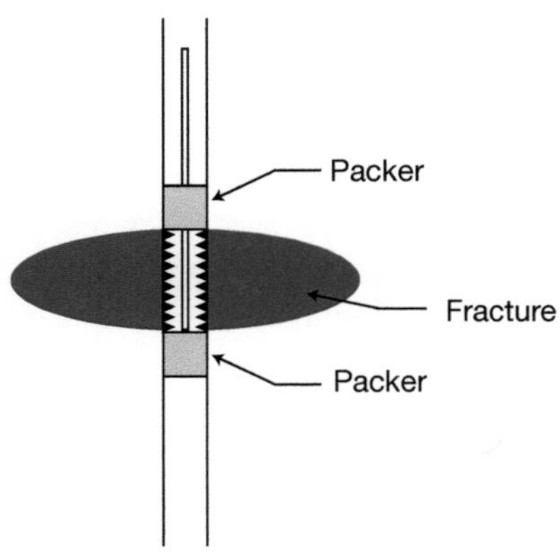

Figure 2.4.1

Schematic diagram.

The following phases occur during hydraulic fracturing:

- initiation of the fracture by pressurising the medium (from the wall of the wellbore or from the perforations),
- extension of the fracture, during which a granular material called a proppant in suspension in the fluid is injected, to keep the fracture open after injection,
- removal of the fracturing fluid and recompletion of the well.

The length of the fractures, typically 50 m to 300 m, depends on the petrophysical and mechanical properties of the rock treated. The volumes injected range from several cubic metres to several thousand cubic metres. Hydraulic fracturing may last from 10 minutes to several hours.

The initiation pressure (also known as the fracturing or breakdown pressure) was discussed in Chapter 1.4.1. Fracturing fluid is first injected with no proppant to widen the fracture, to allow sufficient space for the proppant. In the next phase, the proppant (sand, bauxite, etc.) is mixed with the fracturing fluid which then transports the solids from the wellhead to the tip of the fracture. The fluid viscosity must be high enough to transport the particles but low enough to facilitate the flow. A water- or oil-based gel is generally used, with additives:

- reticulating agent to increase viscosity,
- filtrate reducer,
- breaker to break up the gel after a residence time in the layer.

The main problem when designing a hydraulic fracturing operation is due to the fact that the fluid injected widens and extends the fracture but also leaks off into the formation. These two aspects must therefore be taken into consideration during the calculation.

The fracture geometry depends on the mechanical characteristics of the surrounding rock, its stress state and the fluid used.

2.4.2 Opening a fracture around a vertical wellbore

A fracture around a wellbore only opens if the fluid pressure applied exceeds a certain threshold. The initiation pressure depends on [Sarda, 1984]:

- the mechanical parameters in the overburden, sideburden and underburden of the zone concerned,
- the rheological behaviour of the rock,
- the fluid injection mode,
- leakoff or not.

If we consider elasticity, the stresses present around a vertical wellbore can be calculated (see Chapter 2.2.3). To create a fracture, the fluid pressure at wellbore wall must be greater than the local minimum tangential compressive stress [Boutéca and Sarda, 1987]. Taking the tensile strength T (and tensile failure) as fracture initialisation criterion, the fracture is initiated as soon as the fluid pressure P_f:

$P_f > 3\sigma_3 - \sigma_2 + T$ in impermeable or at least "non-draining" medium,

$P_f - P_R > \dfrac{\nu}{1+\nu}(3\sigma_3 - \sigma_2 + T)$ in permeable medium (where P_R is the reservoir pressure and ν is the rock Poisson's ratio).

For more complete poroelasticity formulations, refer to Charlez (1997).

Studies of failure mechanics demonstrate that discontinuities in a solid produce stress concentrations. At the tip of a fracture the stresses are singular and characterised by a stress intensity factor K_I proportional to the pressure in the fracture and the square root of the fracture dimension. According to a key postulate in linear elastic fracture mechanics, for a given material there exists a critical value of the stress intensity factor $K_{IC,}$ called the fracture tenacity. A fracture will propagate if K_I is greater than or equal to K_{IC}.

The fracture tenacity K_{IC} is a property of a material which expresses the ability of this material to propagate a fracture. The condition at the tip of the fracture must be satisfied:

$$K_I \geq K_{IC}$$

$$\text{with } K_I = \left(p_f - \sigma_{min}\right)\sqrt{\pi l}$$

The tenacity is related to the tensile strength T by the relation $K_{IC} = T\sqrt{\pi l}$, where l is the half-length of an existing crack.

Mini-frac tests involve initiating a hydraulic fracture within a short packed-off interval by injecting low-viscosity fluids slowly through perforated casing or in open hole (see Chapter 1.4.1 and Figure 2.4.2). They represent a reliable method of determining:

- the breakdown pressure,
- the fracture propagation pressure,
- the Instantaneous Shut In Pressure (ISIP),
- the leakoff coefficient.

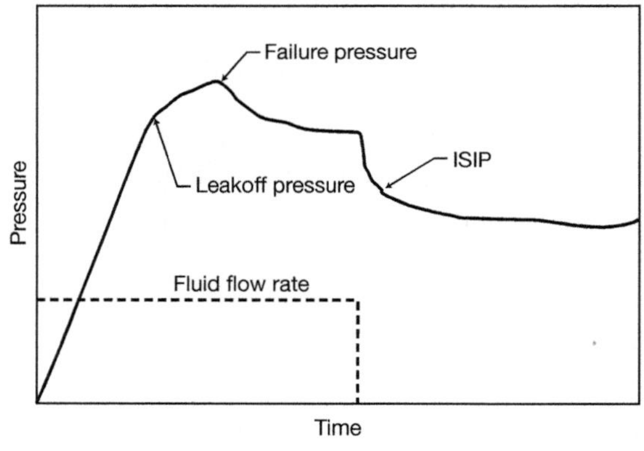

Figure 2.4.2

Pressure variation observed during a hydraulic fracturing test.

Mini-frac tests can be conducted to measure the *in situ* stresses before a fracture stimulation treatment. Carried out both on the reservoir and the surrounding formations, they improve the prediction results.

2.4.3 Description of fractures

Fractures develop and propagate in directions of minimum energy, perpendicular to the minor principal stress (Figure 2.4.3).

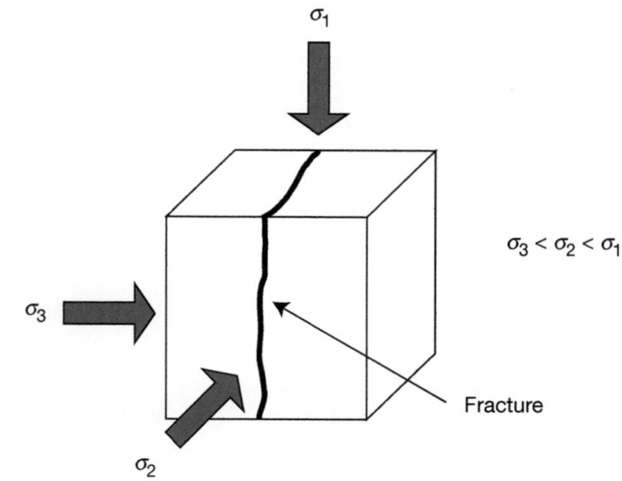

Figure 2.4.3

Fracture and *in situ* stress field.

Since the minor principal stress is often horizontal, the fractures will be mainly vertical. At low depth, however, or in a tectonic context, their dip may be different. If the horizontal stress field is isotropic, the direction of the fractures will be undetermined. Theoretically, in a homogeneous stress field and in a homogeneous isotropic material, a fracture from a point will grow uniformly in all directions to form an almost circular fracture. Initiated in a perforated well, two wings will tend to develop towards the upper part (stresses increasing with depth). Since the reservoir is not infinite, the fracture will eventually reach the upper or lower layers (surrounding formations) which have different mechanical and hydraulic properties. If the minor principal stress in the surrounding formations is greater than that of the reservoir layer (this is the case if it is depleted), the fracture will theoretically be confined within the layer. This aspect is detailed by Charlez (1997). Recent progress in fracture monitoring (bottomhole inclinometry, microseismicity) has demonstrated, however, that contrasts in mechanical properties or stresses cannot always fully explain confinement [Wright *et al.*, 1999].

When hydraulic fracturing is carried out from deviated wellbores, the fractures no longer lie completely in a plane and reoriented multiple fractures may be observed (Figure 2.4.4).

In fractured environments, a complex mixture of natural fractures and initiated fractures is therefore produced.

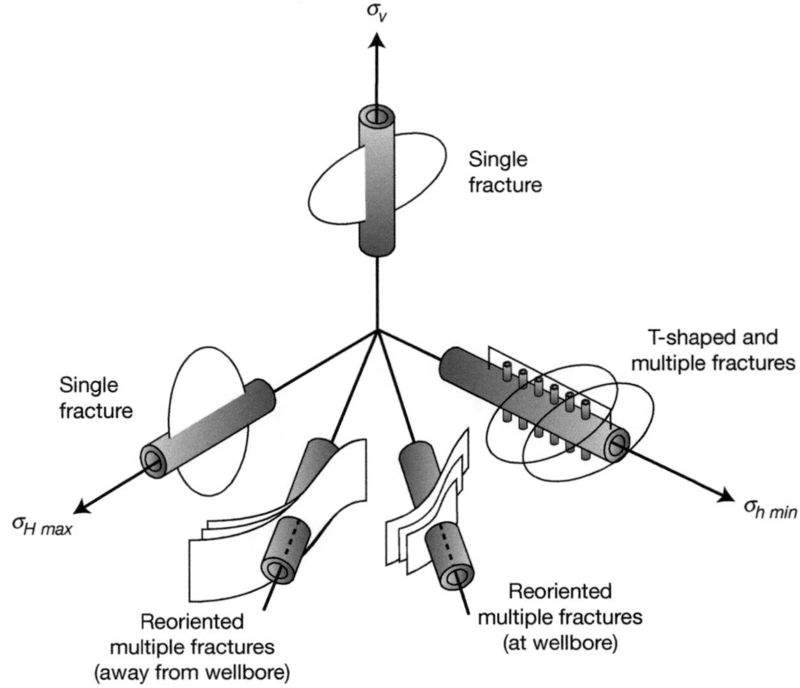

Figure 2.4.4

Fracture geometries as a function of stress field and wellbore inclination [Abass *et al.*, 1995].

2.4.4 Modelling of hydraulic fracture propagation

The calculation methods all simplify the real problem in order to tackle it. The simulation must be able to estimate:

- possible penetration of the fracture in the surrounding formations,
- the proppant injection diagram,
- the final fracture dimensions (height, thickness, length).

The models combine elasticity, fluid flow, mass equilibrium and in some cases a propagation criterion. The models can be classified successively as follows [Mandelsohn *et al.*, 1984]:

- simplified 2D models,
- pseudo-3D models in plane strain,
- fully-3D models.

2.4.4.1 Simplified 2D model: Carter's model

Carter's model, which is the oldest, makes the following assumptions:

- laminar fluid flow in the fracture,
- flow rate maintained constant. The flow rate includes a part $Q1$ allowing extension of the fracture, and a fraction $Q2$ leaking into the formation, at each instant, $Q = Q1 + Q2 = \text{constant}$,
- constant fracture thickness w,
- uniform pressure in the fracture,
- the fluid leaks into the walls of the fracture with a velocity $V(x,t) = \dfrac{C_w}{\sqrt{t - \tau(x)}}$

with C_w the fluid loss coefficient and $\tau(x)$ the time at which the propagating tip of the fracture reaches a location x in the formation.

Then $(t - \tau(x))$ is the elapsed time during which fluid leaked off into the rock formation since its first exposure to the fracturing fluid.

$$C_w = \sqrt{\frac{\sqrt{k\phi C_t}}{\mu}(P_f - P_r)}$$

k permeability of the layer, ϕ porosity, μ viscosity, C_t coefficient
P_f pressure in the fracture, P_r reservoir pressure.

The fracture area is deduced from the flow rate measurements.

$$A = \frac{Qw}{4\pi C_w^2}\left[\exp(\tau_D)erfc(\sqrt{\tau_D}) + \frac{2}{\sqrt{\pi}}\sqrt{\tau_D} - 1\right]$$

$$\text{with } \tau_D = \frac{4\pi C_w^2}{w^2}t$$

Carter's leak-off law is used in the later models.

2.4.4.2 GDK (Geertsma and de Klerk) type 2D models

The problem is considered in 2D in the horizontal plane (x, y). The fracture is assumed to be limited by two horizontal planes separated by height h, constant over time. The fracture therefore has a rectangular cross-section (Figure 2.4.5). The underlying physical assumption is that the lateral edges of the fracture slide freely at the top and bottom of the layer. The width is a function of x.

Geertsma and de Klerk solved the problem analytically.

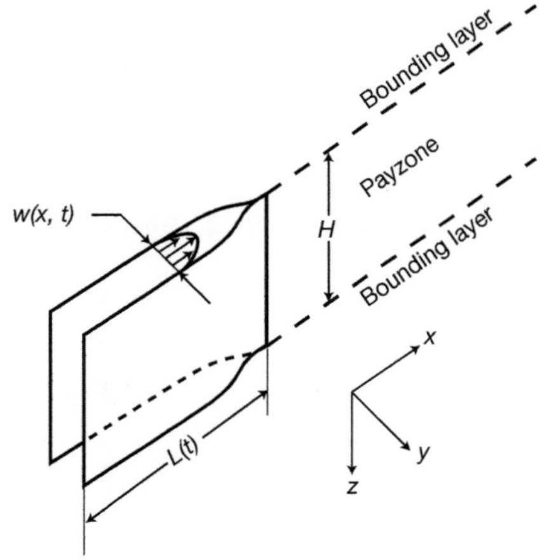

Figure 2.4.5

GDK model [Mendelsohn, 1984].

Charlez gives the following expression:

$$\frac{Q}{h} = \frac{3\pi}{8} w \frac{dL}{dt} + 2C_w \int_0^t \frac{\partial L}{\partial \tau} \frac{d\tau}{\sqrt{t - \tau(x)}}$$

with

h: fracture height,

Q: flow rate,

w: fracture width,

L: total fracture length (1 wing).

For an impermeable rock ($C_w = 0$), we obtain the following expression

$$L(t) = b \left[\frac{GQ^3}{(1-\nu)\mu h^3} \right]^{\frac{1}{6}} t^{\frac{2}{3}}$$

with

$b = 0.68$ for a fracture with one wing and $b = 0.48$ for a fracture with two wings [Meyer, 1986],

G: rock shear modulus,

μ: fluid viscosity,

ν: Poisson's ratio,

and a pressure decreasing over time.

$$P = \sigma_{min} + 7 \left(\mu (1 - \nu) \frac{G^2}{t} \right)^{\frac{1}{3}}$$

According to this analysis, the pressure decreases during fracturing and tends, for large injected volumes, towards the minimum stress.

Since the pressure is considered to be uniform in the fracture, the fracture has an elliptical horizontal cross-section, hence the thickness $w(x)$ at a distance x from the wellbore is given by:

$$w^2(x) = w_0^2\left(1 - \frac{x^2}{L^2}\right)$$

The thickness w_0 at the wellbore is given by

$$w_0 = 2.1\left(\frac{\mu Q L^2}{Gh}\right)^{\frac{1}{4}}$$

In case of leakoff,

we obtain: $$L = \frac{Q}{32\pi h C_w^2}\left(\pi w + 8\sigma_{min}\right)\left(\frac{2\alpha}{\sqrt{\pi}} - 1 + e^{\alpha^2}\,erfc\alpha\right)$$

with $$\alpha = \frac{8C\sqrt{\pi t}}{\pi w + 8\sigma_{min}}$$ and $$w = \frac{3}{2}w_0 = 2.1\left(\frac{\mu Q L^2}{Gh}\right)^{\frac{1}{4}}$$

Length and thickness must be solved simultaneously. Successive approximations are made. A numerical approach is used.

In case of significant leakoff, we obtain: $L = \frac{Q}{\pi C_W h}\sqrt{t}$

2.4.4.3 PKN (Perkins, Kern, Nordgren) model

The model assumes that the plane strain condition is valid in each vertical plane normal to the propagation direction. The pressure is assumed to be constant vertically but it decreases with distance x and is zero at the tip of the fracture. The fracture cross-section is elliptical (Figure 2.4.6). The following analytical solutions are obtained quite simply by associating with these assumptions:

- the 2D expression of fracture thickness,
- the head loss in each fracture length element dL.

Without leakoff, we have:

$$L(t) = C\left[\frac{GQ^3}{(1-\nu)\mu h^4}\right]^{\frac{1}{5}} t^{\frac{4}{5}}$$

with C = 0.68 for 1 wing, 0.45 for 2 wings [Geertsma, 1989]

$$W_0(t) = \left[\frac{(1-\nu)\mu Q^2}{Gh}\right]^{\frac{1}{5}} t^{\frac{1}{5}}$$

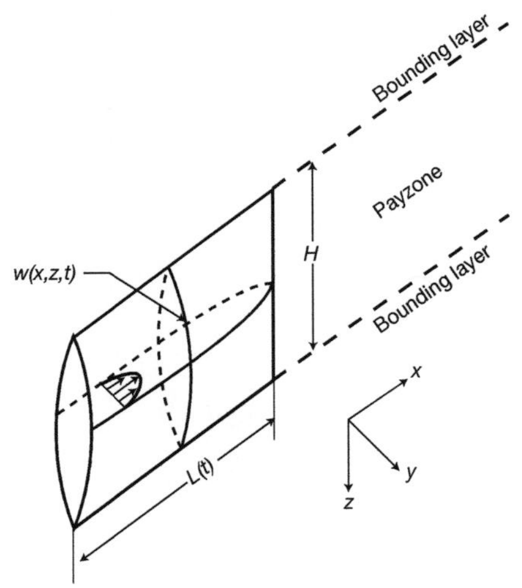

Figure 2.4.6

PKN model [Mendelsohn, 1984].

$$p_0(t)-\sigma_{min}=2\left[\frac{G^4\mu Q^2}{(1-\nu)^4 h^6}\right]^{\frac{1}{5}} t^{\frac{1}{5}}$$

In case of significant leakoff, we obtain:

$$L=\frac{Q}{\pi C_w h}\sqrt{t}\text{, result identical to the GDK approach}$$

$$\text{and } w_0(t)=4\left[\frac{(1-\nu)\mu Q^2}{2\pi^3 G C_W h}\right]^{\frac{1}{4}} t^{\frac{1}{8}} \text{ and } p_0(t)=\sigma_{min}+4\left[\frac{G^3\mu Q^2}{2\pi^3(1-\nu)^3 C_W h^5}\right]^{\frac{1}{4}} t^{\frac{1}{8}}$$

The pressure increases during injection, with an exponent between 1/8 and 1/5.

The intermediate situation between no leakoff and high leakoff is processed numerically.

2.4.4.4 Comments on 2D models

Based on the experience acquired in hydraulic fracturing, we are now in a position to make a value judgement on the models described above.

Carter's model (2.4.4.1) is now of only historical interest.

The GDK model (2.4.4.2) is not suitable for the large majority of cases encountered in hydraulic fracturing: slipping at the limits between reservoir and surrounding formations is the exception rather than the rule.

Comparing the predictions of the PKN model (2.4.4.3) either with more complex models (pseudo-3D, 3D) or with measurements on field, especially pressure measurements, demonstrates surprising efficiency if we consider the simplicity of the assumptions on which it is based [Warpinski *et al.*, 1994].

A key consequence of the PKN model is that it predicts a pressure increase during fracturing if, however, the fracture is reasonably well contained between the surrounding formations. This prediction has led to major progress in the hydraulic fracturing process [Nolte, 1979]: monitoring of the operation by precise pressure measurement (see 2.4.5.1).

2.4.4.5 Pseudo-3D models

2D models cannot simulate vertical and lateral propagation at the same time. Pseudo-3D models have been formulated from the PKN model to manage the variation in height as a function of the distance to the wellbore and the time. The cross-section is no longer elliptical.

These models can be divided into two categories: models which divide the fracture longitudinally into cells whose geometry depends on the fluid pressure and models which use a parametric representation of the total fracture geometry. Each category therefore exhibits different fracture geometries [Warpinski *et al.*, 1994].

Most industrial codes fall into this category.

2.4.4.6 Fully-3D models

Pseudo-3D models cannot process fractures of different shape and orientation, such as those generated from deviated wellbores. Fully-3D models make the fewest initial assumptions. The geometry and topology of the fracture change with the necessarily incremental simulation process [Boutéca, 1988, 1989 ; Carter, 2000; Yamamoto *et al.*, 2000]. No model is perfect, the best is probably the one which is thoroughly understood. The user's experience is vitally important.

Pseudo-3D and fully-3D models have been validated over the years by laboratory experiments.

2.4.5 Fracture monitoring

2.4.5.1 Pressure analysis during fracturing

Pressure analysis is based on information obtained using the PKN model.

If the fracture develops while remaining contained in the layer, the driving pressure $p-\sigma_{\min}$ increases as a function of t^{α} where α is less than 0.2, increase shown by a positive slope α in logarithmic coordinates.

A decrease in α is interpreted as an increase in the fracture height, either in the layer itself or outside it.

If the pressure no longer varies (α equal zero), the situation is considered to be critical. If the subsequent pressure decreases, the most likely cause is vertical extension of the fracture. If the subsequent pressure increases, the cause could be a significant flow restriction in the fracture with proppant jamming [Nolte and Smith, 1981].

2.4.5.2 Geometry of the hydraulic fracture

Numerous techniques can be used to determine the geometry of fractures beyond the wellbore, i.e. the direction, length and even the height [Cipolla and Wright, 2002]:

- surface tiltmeters and downhole tiltmeters in offset well (Figure 2.4.7), resolution decreases with distance,
- induced microseismicity (by far the most frequently used technique) (Figures 2.4.8 and 2.4.9),
- borehole imaging, visualisation of orientation with respect to the borehole.
- radioactive tracers, only near the borehole,
- temperature, only near the borehole.

Passive Seismic Monitoring produces an image of the fracture based on detection and location of the microseisms induced mainly by shear between layers or at naturally closed fractures adjacent to the hydraulic fracture. The sensors used are multicomponent geophones deployed on cable in the well treated or in an observation well.

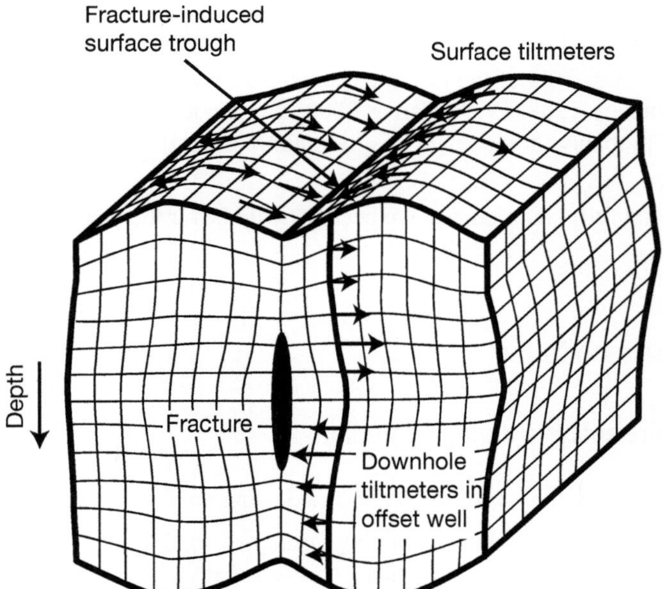

Figure 2.4.7

Principle of tiltmeter measurements in the well [Cipolla and Wright, 2002].

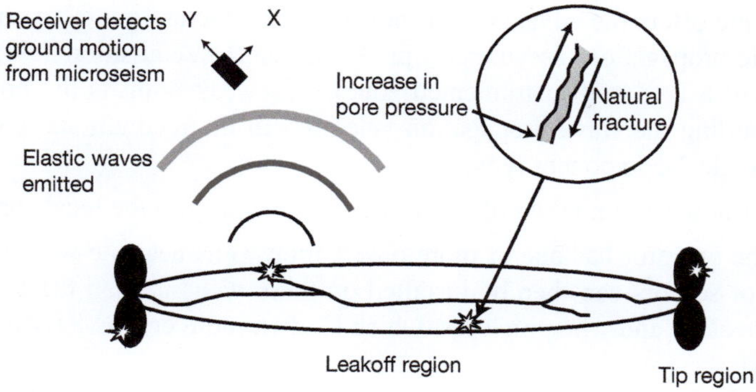

Figure 2.4.8

Principle of hydraulic fracture mapping by microseismicity.

Actual microseismic events from a six-stage hydraulic fracture job.
Each stage is denoted by a different color.

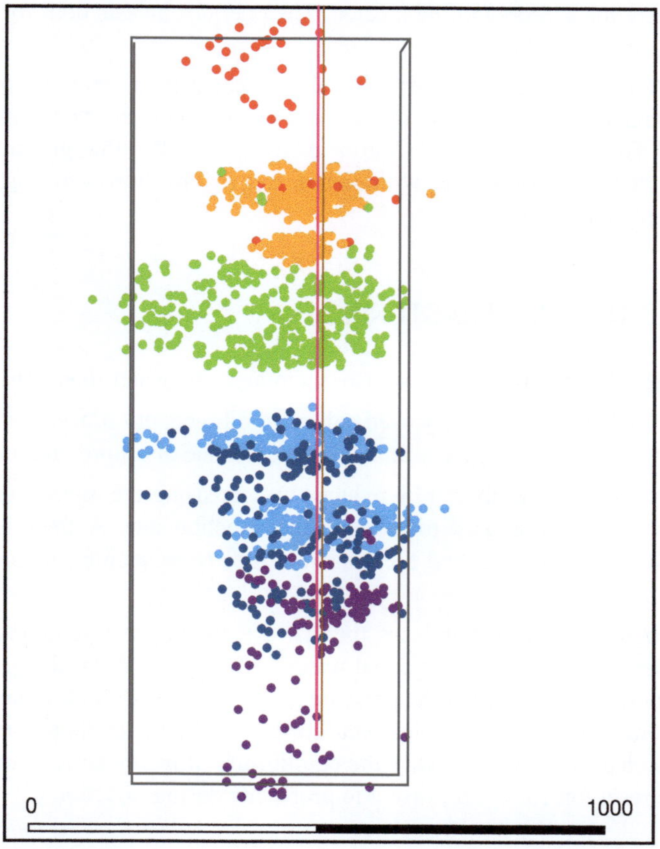

Figure 2.4.9

Microseismicity resulting from hydraulic fracturing [Schlumberger, 2006].

This technique offers the advantage of indicating the fracture geometry and taking into account possible propagation dissymmetry provided that the acquisition device is unbiased (e.g. existence of a less well instrumented zone). The weak point is the possible lack of knowledge regarding the wave propagation velocities in the medium and too few sensors (sometimes a single 3-component sensor).

The operator now has a number of possibilities, depending on the local conditions:

– *Case 1*: the operator has one or more observation wells near the well to be fractured. Several levels of sensors can then be installed on these wells to map the evolution of the fracture as it develops and when it closes, which is often more emissive (two planes sliding over each other).

– *Case 2*: the operator only has the well to be fractured within a radius of several hundred metres. There are two possible configurations depending on whether completion is being installed/restored or already equipped.

If well installation is in process, permanent seismic sensors can be positioned (attached to the production column and coupled to the casing), making sure that cementation is good in the zones where the sensors will be located. The operator invests both for the short-term application and for the long term.

If the well is already equipped, the only solution consists in feeding a probe down the production column and the packer in order to position it against the casing in the producing part of the well. To date, the SIMFRAC probe developed by IFP Energies nouvelles can be used to work under these conditions, with certain restrictions, however, regarding well trajectory and completion.

2.5 INDUCED THERMAL FRACTURING

This chapter is restricted to the fracturing aspects induced by water injection.

Water injection is the most widely used enhanced oil recovery process worldwide. It has a twofold objective: improve recovery and delay the decline of oil production.

The number of injector wells can be reduced by increasing the injection capacity. Fracturing is one of the methods used to increase the injection rate. A fracture increases the exchange area between the well and the reservoir and, consequently, favours higher flow rate for a given pressure.

Water injection in reservoirs induces thermal effects due to a temperature difference between the reservoir rock and the injected water (colder), and hydraulic effects due to an increase in the pore pressure and in the pressure gradient at the well. The pressure and temperature variations affect the mechanical equilibrium of the materials around the wellbore (thermohydromechanical effect), modify the equilibrium of *in situ* stresses and may initiate a fracture in the reservoir rock and cause it to propagate during injection.

2.5.1 Description of fracturing induced by injection of cold fluid

Water injection in reservoirs modifies the thermal and hydraulic equilibria around the wellbore. During injection, the thermal exchanges which take place between the surface and the bottom of the injector well are generally insufficient to raise the temperature of the injected water up to the temperature of the rock at the injection point. Consequently, water is often injected into the reservoir at a temperature rather less than that of the reservoir, which progressively decreases the rock temperature around the wellbore.

The orthoradial stress variation at the wellbore wall due to thermal effects is very difficult to calculate [Charlez, 1996]. Assuming homogeneous material with elastic behaviour, the orthoradial stress variation $\Delta\sigma_\theta$ caused by the temperature difference $\left(T-T_i\right)$ can be expressed as follows:

$$\Delta\sigma_\theta = \frac{E\alpha}{1-\nu}\left(T-T_i\right)$$

with:

- σ_θ tangential horizontal stress at the wellbore wall (orthoradial stress),
- T_i reservoir rock temperature,
- T cooled rock temperature,
- E and ν rock Young's modulus and Poisson's ratio,
- α coefficient of thermal expansion of rock.

The stiffer the rock is (high values of modulus E), the greater the stress variation related to the thermal effects.

For example, for a temperature difference of $\left(T-T_i\right)=1°\ C$, a quick evaluation provides the following thermal stress reduction:

– with a stiff limestone ($E = 50\ GPa$, $\upsilon = 0.2$ and $\alpha = 10^{-5}$/°C), the stress variation $\Delta\sigma_\theta = \dfrac{5.10^5}{0.8}Pa = 0.62 MPa$,

– for a highly porous sandstone ($E = 5\ GPa$, $\upsilon = 0.3$ and $\alpha = 2\ 10^{-6}$/°C), the stress variation $\Delta\sigma_\theta = \dfrac{1.10^4}{0.7}Pa = 0.014 MPa$.

During injection, the decrease in reservoir rock temperature around the injection zone reduces the horizontal stress roughly in proportion to $\dfrac{E\alpha}{1-\nu}\left(T-T_i\right)$, which may favour hydraulic fracturing even though the injection conditions (flow rate and pressure) remain unchanged (constant pressure gradient).

However, water injection also induces an increase in the pore pressure and the pressure gradient at the wellbore. These pressure variations disturb the mechanical equilibrium of the materials around the wellbore and modify the *in situ* stresses, theoretically an increase in the total stresses by poroelastic effect. The two effects are antagonistic [Chavez, 2005].

The thermal effect may prevail in some cases. Field measurements indicate that fracture initiation occurred at a pressure less than the fracturing pressure measured before water injection [Charlez, 1996]. In fracturing regime, leakoff within the fracture extends the zone affected by cooling and therefore favours propagation of the fracture.

During water injection, the presence of solid particles or oil droplets may impair the flow properties of the reservoir rock (hydraulic damage) due to progressive fracture wall plugging. Two plugging modes, which depend on the characteristics of the porous medium and the injection conditions, can be identified. The first is characterised by the formation of a dominant filter cake. This "external" plugging may prevail if poor quality water is injected (high particle concentration and/or large particles). The second is characterised by "diffusion" of solid particles within the medium resulting in "internal" plugging over a depth greater than in the first case. This plugging mode may be encountered if water with low particle concentration and/or small particles is injected. The two plugging modes may coexist. This reduction of the flow properties must be taken into account when studying fracturing by water injection since it has a direct influence on the thermohydromechanical equilibrium of the fracture and therefore the injectivity index.

2.5.2 Modelling

Modelling of Thermal Induced Fracturing (TIF) was initiated by Perkins and Gonzalès (1984, 1985). The PKN model assumes that the fracture is vertical and limited to the height of the reservoir layer, which is nearly always the case. This model is appropriate when the ratio between fracture length and fracture height is much greater than one. The cold zone extends along the fracture with an elliptical shape of half major axis *a* and half minor axis *b* (Figure 2.5.1).

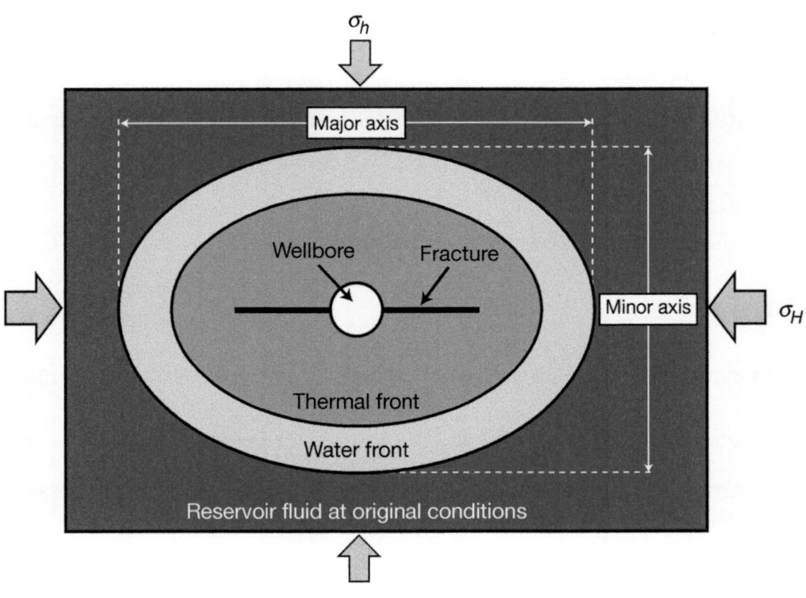

Figure 2.5.1

Cooled zone around the wellbore [Chavez *et al.*, 2005].

The values of a and b are given by the following formulae

$$a = \frac{L}{2}\left(\sqrt{X} + \frac{1}{\sqrt{X}}\right) \text{ and } b = \frac{L}{2}\left(\sqrt{X} - \frac{1}{\sqrt{X}}\right)$$

with
$$X = \frac{2V_c}{\pi L^2 h} + \frac{1}{2}\sqrt{\left(\frac{4V_c}{\pi L^2 h}\right)^2 + 4}$$

where h and L are respectively the fracture height and length and Vc the volume of the cooled zone.

If we assume that no heat is transferred by conduction with the upper and lower bounding layers, the volume of the cold zone Vc is written

$$V_c = \frac{\rho_w C_w W_i}{\rho_r C_r (1-\phi) + \rho_w C_w \phi S_{wi} + \rho_o C_o \phi S_{or}}$$

C_r, C_w and C_o: specific heats of rock, water and oil,
ϕ: rock porosity,
S_{wi}: initial water saturation,
S_{or}: residual oil saturation.

Perkins and Gonzalès (1985) gave an explicit approximation of the orthoradial stress variation:

$$\Delta\sigma_\theta = \frac{E\alpha}{1-\nu}\left(T - T_i\right)$$

$$\left[\frac{b/a}{1+(b/a)} + \left[\frac{1}{1+(b/a)}\right]\left(1/\left\{1+1/2\left[1.45\left(\frac{h}{2b}\right)^{0.9} + 0.35\left(\frac{h}{2b}\right)^2\right]\left[1+\left(\frac{b}{a}\right)^{0.774}\right]\right\}\right)\right]$$

While highly useful for a first estimation of the fracture length, it cannot be used to reproduce complex injection diagrams (two-phase flow, change in mean pressure, etc.).

IFP Energies nouvelles developed the Hydfrac model, based on a based on a 3D two-phase thermal reservoir simulator for heterogeneous and anisotropic porous materials [Charlez, 1996, Longuemare, 2001]. A cold fluid of temperature Tw and flow rate Q is injected into a well. At any time, for a given fracture length the reservoir model calculates the pressure and temperature fields and the fluid saturations. These pressure and temperature variations disturb the mechanical equilibrium and induce stress variations. The stress variations are calculated with a thermoporomechanical model. The minor stresses can be decreased to values below the injection pressure, which results in propagation of the fracture. The fracture description is dynamic (the fracture length may increase or decrease at each time step) and is based on the PKN model.

For each cell fractured, the effect of the damage is represented by a change in transmissivity between fracture and reservoir.

The injection conditions rarely remain constant during the life of the injector well. In particular, water injection may be stopped for one or more days. One of the main problems associated with these periods during which injection is stopped is how to determine the change in fracture length during the period. Models show that the fracture closure rate depends on the injection duration period prior to shut-in [Longuemare *et al.*, 2001] (Figure 2.5.2).

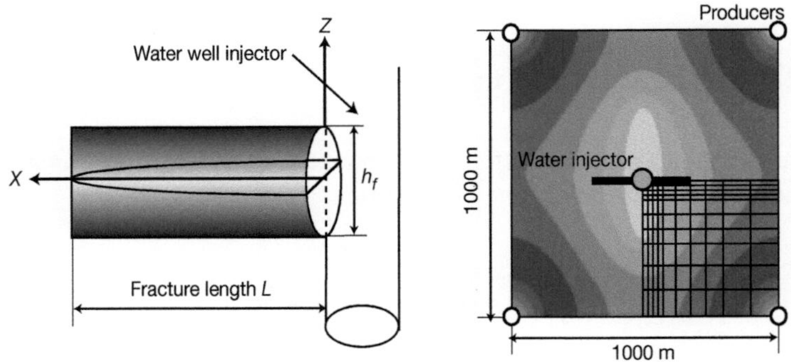

Figure 2.5.2

PKN fracture model and simulation [Longuemare *et al.*, 2001].

Simulations have been carried out using data from a field located in south-west France in a carbonate reservoir at a depth of 3,000 m. The cold fluid flow rate is an input and the bottomhole pressure is deduced from injectivity index calculations. Given the highly complex conditions, the agreement between measurement and Hydfrac simulation results is excellent. The thermo-poromechanical effects, fracture propagation and pressures in the fractures seem to be properly simulated (Figure 2.5.3.).

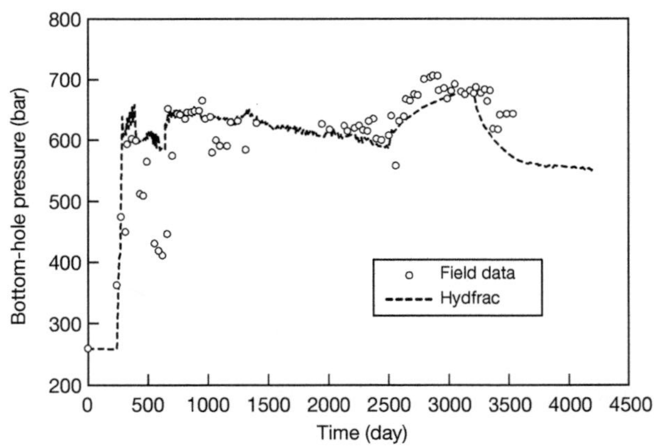

Figure 2.5.3

Calculated and measured bottom-hole pressures [Field data are taken from (Détienne *et al.*, 2002)].

3

Geomechanics and Reservoir

It has become standard practice in reservoir engineering to use a geocellular 3D model for the reservoir and the surrounding rocks in order to study the possible combined effects between fluid production and the mechanical strains of the structure. It is in fact very important to predict possible subsidence of the surface as oil and gas production proceeds.

Even a slight variation in pore pressure in a reservoir may cause significant displacements in the subsurface layers, especially if the reservoir is very poorly or weakly cemented, which increases the risk of collapse and the development of shear zones. Combined problems such as these are encountered when studying the stability of CO_2 sequestration sites, where it is important to check that the injection of large quantities of gas into the structure will not destroy its long-term mechanical stability.

Exploitation of an oil reservoir is very expensive and entails a high degree of uncertainty. These costs and uncertainties can be reduced by reservoir studies conducted by the oil companies before or during exploitation. These reservoir studies are generally based on numerical simulations of flows in the reservoir in order to predict the quantities of hydrocarbons present and their production over time, optimise reservoir production and assess the risks associated with production. Reservoir simulators give a highly accurate description of the fluid and heat transfers in the reservoir but do not model (or do so in a very simplified manner only) the mechanical behaviour of the reservoir during its production. Production in certain types of reservoir (poorly consolidated reservoirs, fractured or faulted reservoirs, etc.) is accompanied by major mechanical effects likely to threaten the integrity of the reservoir and its wells, generate subsidence of the surface layers and modify the flows of the reservoir fluids, and therefore any production predictions made by the reservoir engineers. Modelling these mechanical effects therefore requires a geomechanical model which must be coupled with the reservoir simulation.

We will first describe the hydraulic diffusivity equation, a classical model of the single-phase and isothermal flows in porous media. This diffusivity equation involves two important parameters: the compressibility and the permeability reservoir rock. We will then demonstrate that these two parameters may be highly dependent on the stress state in the reservoir. In this case, it becomes essential to model the mechanical behaviour of the oil reservoir during production. We then describe how the coupling between the reservoir and geomechanical models can be achieved. Two examples of coupling between reservoir and geomechanical simulators are described for a high porosity reservoir and a faulted reservoir. Lastly, we will study the evolution of stresses in the reservoir during depletion and the measurements available to validate it.

3.1 MODELLING THE GEOMECHANICAL AND RESERVOIR COUPLING

3.1.1 Flow and permeability

3.1.1.1 Hydraulic diffusivity equation

The phenomena involved during production of oil or gas from a reservoir and when injecting fluids (water, gas, etc.) to enhance production are mainly related to the flow of fluids in porous media (see Bear and Buchlin, 1991, for example). To simplify the description, we will restrict ourselves to single-phase isothermal flows in porous media. We write ϕ the porosity of the porous medium (Eulerian porosity, which is the ratio of the pores to the current total volume) and ρ the density of the saturating fluid. The conservation of fluid mass can then be expressed as:

$$\frac{\partial}{\partial t}(\rho\phi) + \mathrm{div}(\rho\phi v) = 0 \tag{3.1.1}$$

with v the seepage velocity of the fluid in the porous medium. The latter is generally related to the fluid pressure p via Darcy's law which, in the absence of gravity, can be written:

$$\phi v = -\frac{k}{\mu}\,\mathrm{grad}\; p \tag{3.1.2}$$

with k the rock intrinsic permeability and μ the fluid dynamic viscosity.

The fluid is assumed to be compressible, of compressibility C_f defined by:

$$\partial\rho = \rho_0 C_f \partial p \tag{3.1.3}$$

with ρ_0 the density of the fluid under fluid pressure p_0.

A reservoir rock "compressibility" C_r is introduced at the same time, to take into account a variation of the porous medium porosity as a function of the saturating fluid pressure:

$$\partial\phi = \phi_0 C_r \partial p \tag{3.1.4}$$

with ϕ_0 the initial porosity of the porous medium given for the initial fluid pressure p_0. Introducing equations (3.1.2) to (3.1.4) in the conservation of mass equation (3.1.1), we obtain after linearisation:

$$\phi_0(C_f + C_r)\frac{\partial p}{\partial t} - \frac{k}{\mu}\Delta p = 0 \tag{3.1.5}$$

Associated with representative boundary and initial conditions, equation (3.1.5) gives the evolution of saturating fluid pressure in the porous medium. It is important to note, however, that the compressibility C_r is not an intrinsic property of the reservoir rock but depends on the stress path followed by the reservoir during its depletion. In addition, the permeability of the porous medium may vary significantly during depletion of a reservoir due to changes in effective stress in the reservoir.

3.1.1.2 Reservoir rock compressibility

Equation (3.1.5) shows that the evaluation of recovery is based mainly on the estimation of reservoir rock compressibility, which represents the strain of the porous volume induced by the variation in reservoir pressure:

$$C_r = \frac{1}{V_p}\frac{\Delta V_p}{\Delta p_p} \tag{3.1.6}$$

So defined, the reservoir rock compressibility depends on the stress path followed by the reservoir during production. This stress path may be quite different from the isotropic or oedometric loading paths traditionally adopted in the laboratory to evaluate C_r. In particular, it may be heterogeneous and vary during depletion (see Chapter 3.1.3).

Concerning this point, note that the compressibilities defined by Zimmerman describe exclusively the volume response of a rock subjected to hydrostatic loading $\underline{\underline{\sigma}} = -p_c \underline{\underline{1}}$ and to a pore pressure p_p [Zimmerman 1991]:

$$C_{bc} = -\frac{1}{V_b}\frac{\partial V_b}{\partial p_c}\bigg|_{p_p} \qquad C_{pc} = -\frac{1}{V_p}\frac{\partial V_p}{\partial p_c}\bigg|_{p_p}$$

$$C_{bp} = -\frac{1}{V_b}\frac{\partial V_b}{\partial p_p}\bigg|_{p_c} \qquad C_{pp} = -\frac{1}{V_b}\frac{\partial V_b}{\partial p_p}\bigg|_{p_c} \tag{3.1.7}$$

where V_p designates the porous volume and V_b the bulk volume, the compressibility closest to definition (3.1.6) being obviously C_{pp}.

Assuming linear poroelastic behaviour of the reservoir rock, the compressibilities so defined can be expressed as a function of the drained bulk modulus K_o and Biot's coefficient b [Boutéca, 1992]:

$$C_{bc} = \frac{1}{K_0} \qquad C_{pc} = \frac{b}{\phi_0 K_0} = \frac{C_{bp}}{\phi_0}$$

$$C_{bp} = \frac{b}{K_0} \qquad C_{pp} = \frac{b-\phi_0(1-b)}{\phi_0 K_0} \tag{3.1.8}$$

For an incompressible solid matrix (i.e. $b = 1$), we obtain simply:

$$C_{bc} = C_{bp} = \frac{1}{K_0} \qquad C_{pc} = C_{pp} = \frac{1}{\phi_0 K_0} \tag{3.1.9}$$

Still assuming linear poroelastic behaviour of the reservoir rock, we can determine expressions for C_r as a function of the loading path considered. The three types of loading path generally followed to simulate depletion in the laboratory are the hydrostatic, oedometric and proportional paths (see Chapter 3.1.3):

$$C_r^{hydrostatic} = \frac{1}{\phi_o}\left(\frac{b^2}{K_o} + \frac{b-\phi_o}{K_s}\right)$$

$$C_r^{oedometric} = \frac{1}{\phi_o}\left(\frac{b^2}{Ko + \dfrac{4G}{3}} + \frac{b-\phi_o}{K_s}\right) \tag{3.1.10}$$

$$C_r^{proportional} = \frac{1}{\phi_o}\left(\frac{1+2K}{3}\frac{b^2}{K_o} + \frac{b-\phi_o}{K_s}\right)$$

For a hydrostatic loading path, we obtain the expression for C_{pp} expressed in this case as a function of the bulk modulus of the solid matrix.

Table 3.1.1 compares the compressibilities estimated from expressions (3.1.8) and (3.1.10). The poroelastic parameters used are extracted from the thesis of D. Bévillon (2000) and are representative of a highly compactable reservoir chalk. Note that the values obtained largely depend on the assumptions made concerning the loading path.

Table 3.1.1 Estimation of chalk compressibility as a function of the loading path followed.

Assumed stress path	Calculated compressibility (GPa^{-1})
Hydrostatic path (C_{bc})	0.71
Hydrostatic path (C_{pp})	1.76
Oedometric path	0.88
Proportional path (K = 0.5)	1.18

In addition, for some rocks, the compressibility may depend, like the drained bulk modulus, on the pore pressure and stress state [Bemer *et al.*, 2001].

3.1.1.3 Intrinsic permeability variation

The experimental data obtained on various reservoir rocks (sandstones and carbonates) show that the permeability of a rock subjected to compression decreases during the initial nonlinear phase when the existing microcracks close and the linear phase associated with pore elastic deformation. The change in permeability during the post-elastic phase is more complex: examples where the permeability decreases [Zhu and Wong 1997] [Brawn and Jahns 1998] [Yale and Crawford 1998] and examples when the permeability increases [Rhett and Teufel 1992a] [Keaney *et al.* 1998] [Brawn and Jahns 1998] [Yale and Crawford 1998] have been encountered in the laboratory. The results obtained demonstrate that the change in post-elastic permeability depends on the loading path followed and the initial rock porosity and permeability levels.

It is important to note that the various studies mentioned are based on traditional triaxial tests where the stress state is axisymmetric and where the permeability is measured along the axial direction. Microcracks develop preferentially along the axial direction, which is also that of the flow. Consequently, when the authors observe an increase in permeability with the dilatancy of the sample, the permeability is measured parallel to the main axis of the cracks. The petroleum industry, however, is more concerned by the change in horizontal permeability.

Schutjens and de Ruig (1996) used a special cell to measure the change in radial permeability (flow perpendicular to the preferential orientation of the cracks). Sandstone samples with high initial porosity and permeability levels (porosity between 21% and 32%, radial permeability between 0.1 mD and 500 mD) have been tested under hydrostatic and oedometric loading paths. The true depletion tests showed a decrease in radial permeability. However, the test conditions (loading paths and initial permeability levels of the samples) correspond *a priori* to cases where the axial permeability also decreases. By comparing tests with measurement of the axial permeability and tests with measurement of the radial permeability, conducted on the same rock of low initial porosity and along a deviatoric loading path with low mean effective stress, it would be possible to check whether the radial permeability increases in the same way as the axial permeability.

Only very few data are available in the literature on the behaviour of a rock subjected to a real triaxial loading path and on the anisotropy of the permeability tensor. Al-Harthy *et al.* (1998) conducted depletion tests on various sandstones for three different stress states: true triaxial (p_o; $0.8{*}p_o$; $0.6{*}p_o$), triaxial (p_o; $0.8{*}p_o$; $0.8{*}p_o$) and hydrostatic (p_o; p_o; p_o). The results have shown a decrease in the vertical permeability (direction of the major principal stress) with the drop in pressure, this loss of permeability being lower for true triaxial loading than for traditional triaxial loading or hydrostatic loading.

Al-Harthy *et al.* (1999) completed these results by studying, still for sandstones, the anisotropy of the permeability tensor and the influence of the loading path on the change in permeabilities in the three main directions of the true triaxial test. For samples exhibiting horizontal stratification, they measured vertical anisotropy factors (ratios between the vertical permeability and a horizontal permeability) of up to 40; this factor is obviously greater for true triaxial loading ($1.25{*}p_o$; p_o; $0.75{*}p_o$) than for hydrostatic loading (p_o; p_o; p_o). The horizontal anisotropy factors (ratios between the two horizontal permeabilities) measured are more limited and lie between 1.1 and 1.7, but this difference in horizontal permeabilities may prove extremely important when installing a horizontal well. The results obtained do not indicate a clear trend as to the influence of depletion on the anisotropy of permeabilities; a slight decrease in anisotropy with increase in effective stresses has nevertheless been observed on a reservoir sample exhibiting horizontal stratification.

Some authors have tried to define variation laws (mainly empirical) expressing permeability as a function of the stress path followed. The small permeability variations in the elastic range can be estimated using the experimental law proposed by David *et al.* (1994):

$$k = k_o e^{\left[-a\left(p'-p_o'\right)\right]}$$

where k_o is the permeability measured at p_o'.

A variation law was also proposed by Boutéca *et al.* (2000) to describe the change in permeability between the damage and brittle failure criteria in the case where permeability decreases, i.e. for high initial permeabilities. This law involves the stress path followed defined by the derivative of the function $p' = f(q)$ (effective mean stress – deviatoric stress):

$$-\frac{1}{k_0}\frac{\partial k}{\partial q} = a\frac{\partial p'}{\partial q} + b$$

where k_o is the permeability before damage.

For low porosity rocks, permeability is likely to increase as the rock is subjected to shear-induced dilation. This increase may be related to the strains in the rock, the dilating strain in a given direction being associated with the width of the crack created.

In case of ductile failure of the reservoir rock (very unlikely *in situ*), no permeability reduction law is available.

Currently therefore, there is no generic law describing the change in intrinsic permeability of a reservoir rock as a function of the stress path.

3.1.2 Coupling between geomechanical and reservoir models

3.1.2.1 Conventional reservoir approach

Reservoir simulations are based on a reservoir model which describes the reservoir geometry, heterogeneities of petrophysical properties in the reservoir, thermodynamics of the reservoir fluids, flows in porous media, etc. The models used to predict the production of reservoir fluids are more or less complex: single-phase model, Black-Oil model, compositional model, treatment of thermal effects, case of flows in fractured reservoirs.

Traditionally in reservoir engineering, the mechanical behaviour of the reservoir is ignored and modelling focuses on the fluid flows in the reservoir layers. In the simplest case of a single-phase isothermal flow, the reservoir simulator solves an equation similar to the hydraulic diffusivity equation (3.1.5). In this approach, the rock permeability and compressibility are generally constant, which amounts to neglecting the influence of the rock strains and setting *a priori* the stress path followed (generally with uniaxial or isotropic strain) for each mesh of the reservoir model [Boutéca, 1992]. Some simulators nevertheless allow these parameters to vary with the reservoir fluid pressure.

3.1.2.2 Limitations of the conventional reservoir approach

The pressure, temperature and saturation variations generated by exploitation modify the distribution of stresses in the reservoir. The stress variations associated with reservoir production may be extremely high in poorly consolidated reservoirs and fractured and faulted reservoirs. Consequently, the porosity, compressibility and permeability of the reservoir rock are likely to vary while the reservoir is producing. While conventional reservoir simulators assume that these parameters are constant or depend only on the fluid, in actual fact they depend on the stress path followed by the reservoir rock. The pressure dependence of porosity imposed via the rock compressibility introduced in conventional reservoir simulators cannot alone describe the behaviour of the reservoir rock during production [Gutierrez and Lewis, 1998]. To obtain a precise description of the change in reservoir porosity and permeability, the geomechanical behaviour of the reservoir during production must be modelled.

The main geomechanical effects affecting production are the compaction and associated increase in the recoverable quantities of fluid. Obviously, these effects are more noticeable for highly compactable reservoir rocks (poorly cemented sandstones generally qualified as poorly consolidated, highly porous chalks, etc.). Compaction is due to the decrease in reservoir pressure and the associated increase in effective compressive stresses in the reservoir. The mechanical behaviour of poorly consolidated reservoir rocks is complex, being highly nonlinear and depending on the stress path and the temperature. With highly porous chalks, the mechanical behaviour of the reservoir rock may also be considerably affected by variations in water saturation ("water weakening") [Hermansen *et al.*, 2000; Homand, 2000; Matà, 2001]. Compaction phenomena play a significant role in the production of reservoir fluids, as demonstrated by the Bachaquero fields [Merle *et al.*, 1976] and Zuata [Charlez, 1997] in Venezuela. For highly compactable reservoirs, therefore, it is essential to predict the change in reservoir porosity during production using strain or stress data provided by a geomechanical model associated with the reservoir model.

The stress variations resulting from reservoir production modify the porous structure of the rock and therefore its permeability. The permeability of highly consolidated and naturally fractured or faulted reservoirs is extremely sensitive to stress changes during production. The thermoporoelastic effects in these reservoirs may modify the conductivities of the fractures and make waterflooding much less effective [Guttierez & Makurat, 1997]. For these reservoirs, the thermoelastic effects are generally concentrated around injector wells, where the temperature variations are higher, whereas hydromechanical effects affect the entire reservoir. The thermohydromechanical behaviour of the fracture system is complex. Macroscopic behaviour depends on the density and orientation of the fracture system, the initial stress state and the stress path in the reservoir during production. Due to the high elastic rigidity of the matrix (rock intact), the strains are mainly located on the fracture planes, thereby generating a change in their hydraulic conductivity [Koutsabeloulis *et al.*, 1994; Heffer *et al.*, 1994]. Geomechanical modelling of the reservoir is therefore required to estimate the permeability changes in the reservoir, whether they are related to changes in the porous structure or to activation of fractures or faults.

3.1.2.3 Geomechanical approach

Hydrocarbon production modifies the stress and strain fields around the reservoir, so the geomechanical model must therefore take into account the behaviour of the surrounding formations. The reservoir overburden is generally modelled up to ground level, where a free surface condition is imposed. The sideburden and underburden taken into account in the geomechanical model are assumed to be quite extensive so that the zero normal displacement conditions can be imposed at the boundaries of these surfaces. Osorio *et al.* (1998) demonstrated that the conditions at the mechanical limits imposed on the boundaries of the surrounding formations influenced the permeability changes in the reservoir.

To take into account the above-mentioned mechanical effects, the equations of equilibrium in the reservoir and its surrounding formations must be solved, in addition to the hydraulic problem. The equations of this problem can be obtained from the poroelastic constitutive laws proposed by Coussy (1995) for an isotropic material. Using the tensorial notation and the convention of mechanics of continuous media (negative compression), these equations can be written as:

$$\partial \underline{\underline{\sigma}} = (K_0 - \frac{2}{3} G) \partial \underline{\underline{\varepsilon}} \underline{\underline{1}} + 2G \partial \underline{\underline{\varepsilon}} - b \partial p \underline{\underline{1}} \tag{3.1.11}$$

$$\frac{1}{M} \partial p = - b \partial \underline{\underline{\varepsilon}} + \frac{1}{\rho_0} \partial m \tag{3.1.12}$$

with M Biot's modulus, b Biot's coefficient, G the rock shear modulus and K_0 the drained bulk modulus of the rock. $\underline{\underline{\sigma}}$ and $\underline{\underline{\varepsilon}}$ designate respectively the stress and strain tensors, $\underline{\underline{1}}$ the Kronecker symbol, p the pore pressure and m the fluid mass intake with respect to the initial total volume. The first poroelastic constitutive law is introduced in the mechanical equilibrium equation which, in the absence of body force, is written:

$$\text{div } \sigma = 0 \tag{3.1.13}$$

The strain tensor is therefore expressed as a function of the displacement vector $\underline{u}$. The previous equation therefore becomes:

$$G \Delta \underline{\boldsymbol{u}} + (K_0 + \frac{G}{3}) \text{grad } \underline{\boldsymbol{u}} = b \text{ grad } p \tag{3.1.14}$$

In addition, the fluid mass intake m with respect to the fluid initial volume is expressed from the mass balance and Darcy's law as:

$$\frac{\partial m}{\partial t} = -\text{div}\left(\rho \phi v\right) = -\text{div}\left(\rho \frac{k}{\mu} \text{grad } p\right) \tag{3.1.15}$$

The latter equation is introduced in the second poroelastic constitutive law, to give after linearisation:

$$\frac{1}{M} \frac{\partial p}{\partial t} - \frac{k}{\mu} \Delta p = -b \frac{\partial \varepsilon_v}{\partial t} \tag{3.1.16}$$

with $\underline{\underline{\varepsilon}}_v = \text{tr } \underline{\underline{\varepsilon}} = \text{div } \underline{u}$, the volume strain.

The geomechanical problem therefore consists in solving the following system of equations system which describes the flows and elastic strain of the porous medium [see also Boutéca, 1992; Coussy, 1995; Lewis and Schrefler, 1998]:

$$G\Delta\underline{\boldsymbol{u}}+(K_0+\frac{G}{3})\text{grad }\underline{\boldsymbol{u}}=b\text{ grad }p \qquad (3.1.17)$$

$$\frac{1}{M}\frac{\partial p}{\partial t}-\frac{k}{\mu}\Delta p=-b\frac{\partial \varepsilon_v}{\partial t} \qquad (3.1.18)$$

The pore pressure p is the main unknown of equation (3.1.18) and the displacement $\underline{u}$ is the main unknown of equation (3.1.17). The displacement $\underline{u}$ appears in the right member of the "hydraulic" equation (3.1.18) via the volume strain. Similarly, the fluid pressure modifies the mechanical equilibrium via the right member of equation (3.1.17). Equations (3.1.17) and (3.1.18) are therefore coupled.

3.1.2.4 Coupling between fluid flow and geomechanics

To identify the various levels of coupling between the hydraulic and mechanical aspects, we examine the discretised form of the system of equations (3.1.17) and (3.1.18). After discretisation in space and time, the system of equations (3.1.17) and (3.1.18) is written in the following matrix form [Settari and Walters, 1999]:

$$\begin{bmatrix} K & L \\ L^T & E \end{bmatrix}\begin{bmatrix} \Delta_t \underline{\boldsymbol{u}} \\ \Delta_t p \end{bmatrix}=\begin{bmatrix} \underline{\boldsymbol{F}} \\ R \end{bmatrix} \qquad (3.1.19)$$

with:

- $\boldsymbol{u}$ the discretised mechanical unknown,
- p the discretised pressure unknown,
- Δ_t the time difference operator,
- K the mechanical rigidity matrix or stiffness matrix,
- L the coupling matrix between mechanical and hydraulic unknowns,
- $\boldsymbol{F}$ the force vector,
- E the hydraulic flow matrix in porous medium,
- R the source term of the flow problem.

The hydraulic flow matrix E is generally decomposed in the form $E = T - D$, where T is a transmissivity matrix and D the accumulation diagonal matrix. It is important to note that in the case described, matrices K and E are considered as linear operators whereas for a more realistic problem (elastoplastic behaviour, compaction, multiphase flow, etc.), these operators are nonlinear.

Various coupling levels can be considered when solving the matrix system (3.1.19):

- **Conventional reservoir approach:** in this approach, the coupling term in the second equation of matrix equation (3.1.19) is reformulated as a function of the pressure variation so that matrix D uses the rock and fluid compressibilities. The limitations of this approach have already been discussed. It can nevertheless be used to obtain an estimation of the stress variations when the pressure variations calculated by the reservoir simulator are transmitted to the second member of the first equation of matrix equation (3.1.19) via a geomechanical simulator. In this case, we speak of **one-way coupling** or decoupling.

- **Partially coupled:** in this approach, the equations of matrix system (3.1.19) are solved separately on different simulators. A reservoir simulator solves the hydraulic problem and a mechanical simulator the geomechanical problem. Coupling is carried out by exchanging results between the two simulators at different exchange periods. In practice, coupling consists in calculating the pressure field p from the displacement field $\underline{u}$ calculated during the previous geomechanical simulation with the equation:

$$E\Delta_t p = R - L^T \Delta_t \underline{\boldsymbol{u}} \qquad (3.1.20)$$

  This first calculation is carried out with a conventional reservoir simulator using a porosity correction calculated from a mechanical variable [Settari and Mourits 1994, Settari and Mourits 1998, Bévillon 2000, Mainguy and Longuemare 2002]. Then, using the pressure field found using (3.1.15), the displacement vector is calculated from:

$$K\Delta_t \underline{\boldsymbol{u}} = \underline{\boldsymbol{F}} - L\Delta_t p \qquad (3.1.21)$$

  This second calculation is carried out using a geomechanical simulator in order to apply the reservoir model pressure variations as external load. This type of coupling is said to be **explicit** if it is not iterated up to convergence of mechanical and hydraulic unknowns over the same period [Fung *et al.* 1992, Tortike and Farouq Ali, 1993, Settari and Walters, 1999] In contrast, we speak of **iterative coupling** if the process is iterated up to convergence of unknowns [Settari and Mourits 1994, Chin and Thomas, 1999]. The main advantage of loose coupling is that reservoir and geomechanical simulators, already highly developed in their respective domains, can be reused. This method also allows the use of different time and space discretisations for the hydraulic and geomechanical problems.
- Fully coupled: in this approach, equations (3.1.19) are solved simultaneously in the same simulator. The space discretisation method is then the same for both equations and may be either finite difference type [Stone *et al.,* 2000] or finite element type [Lewis and Schrefler 1998, Koutsabeloulis and Hope 1998]. This approach provides a consistent solution to the coupled problem but involves solving all equations at each time step.

3.1.2.5 Example of external coupling in porosity on a highly compactable reservoir

This section illustrates the iterative loose coupling methodology where the porosity injected in the reservoir model is calculated using strains calculated by a geomechanical model. The results presented are extracted from the thesis of D. Bévillon (2000) and obtained from the SARIP^{CH} reservoir simulator and ABAQUS finite element software on a synthetic reservoir case. The example illustrates the depletion of a highly compactable homogeneous reservoir, of simplified geometry with a producing well located in its centre. The geometry of the reservoir and its finite volume mesh are shown on Figure 3.1.1. The finite element mesh used in ABAQUS includes the reservoir and the side, over and underburdens (Figure 3.1.2). The simplified reservoir geometry and the choice of finite volume and finite element discretisations simplify data exchange between the two models.

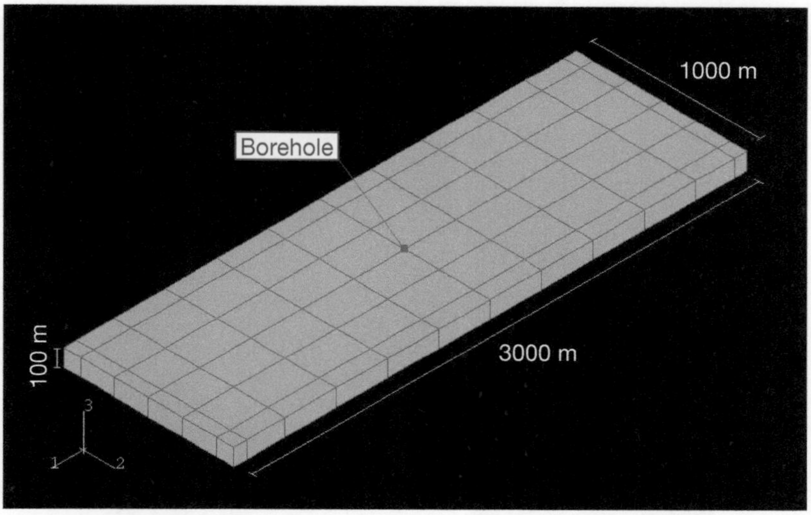

Figure 3.1.1

Reservoir mesh composed of 55 finite volume cells.

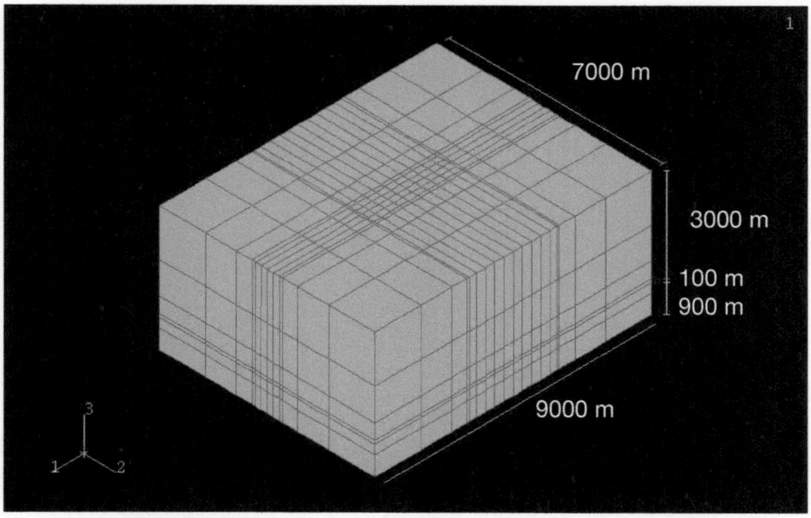

Figure 3.1.2

Geomechanical mesh composed of 1296 Q20 finite elements.

The reservoir pressure is initially 48 MPa and the reservoir is produced from the central producing well whose bottomhole pressure is fixed at 10 MPa. The reservoir fluid is assumed to be poorly compressible. The reservoir porosity is initially high (40%) and the mechanical behaviour is assumed to be nonlinear and represented by a Cam-Clay model. The behaviour of the surrounding formations is represented by a linear poroelastic model which provides the changes in pressure and stress field. Porosity and permeability are low in the surrounding formations.

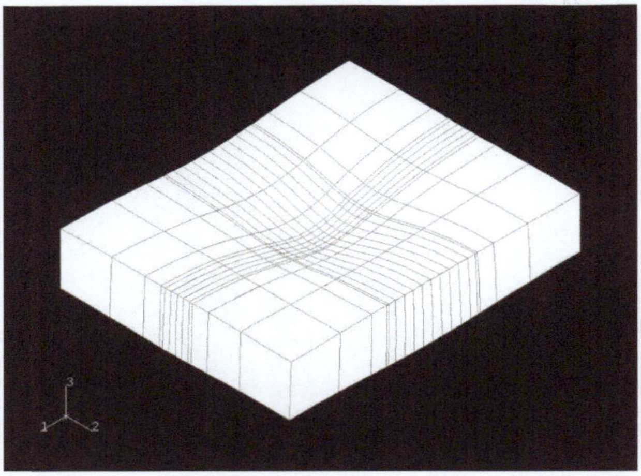

Figure 3.1.3

Deforming finite element mesh after 33 years production.

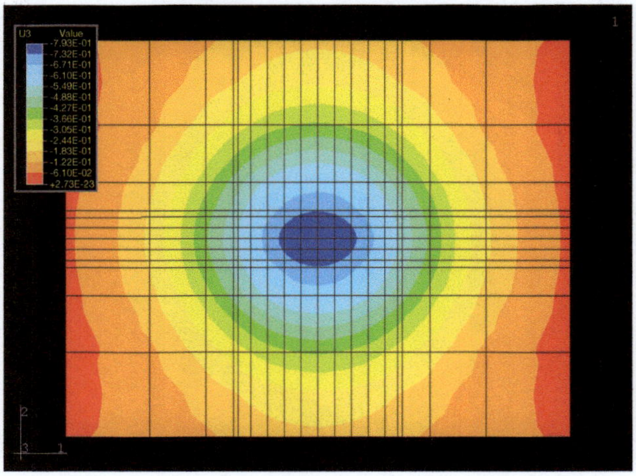

Figure 3.1.4

Vertical displacement isovalues on the surface after 33 years production.

The results of the coupled simulations indicate compaction of the reservoir rock resulting as surface subsidence (Figures 3.1.3 and 3.1.4). After 33 years production, the vertical displacement reaches a maximum value of 75 cm on the surface and in the centre of the reservoir. The hydromechanical calculations provide the change in pore pressure in the surrounding formations. These results indicate overpressure in the area of the surrounding formations located at the periphery of the reservoir, reaching 7 MPa afters 12 years production (Figure 3.1.5). This overpressure is due to stress transfer from the reservoir to the area of the surrounding formations in contact with the reservoir, since their rigidity is higher than that of the reservoir. This phenomenon is generally called "arch effect" or "stress arching", [Osorio *et al.,* 1998]. This stress transfer increases the pore pressure near the reservoir, which cannot be dissipated due to the low permeability of the surrounding formations.

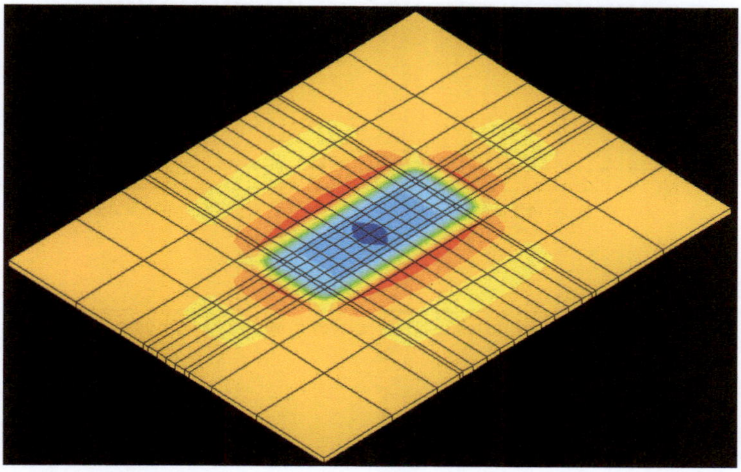

Figure 3.1.5

Pore pressure isovalues in the reservoir and sideburdens after 12 years production.

It is worthwhile comparing coupled geomechanical-reservoir simulations with conventional reservoir simulations to determine the impact of the reservoir mechanical behaviour on production. The previous results were therefore compared with conventional reservoir simulations for which various reservoir rock compressibility values were used. The variations laws of compressibility as a function of pore pressure which have been tested are shown on Figure 3.1.6. The "*Cp* elastic" and "*Cp* oedometric" compressibilities are obtained for an oedometric loading path assuming that the reservoir rock behaviour is respectively elastic and elastoplastic (modified Cam-Clay model), the "*Cp* mean compressibility" corresponding to the mean value between the maximum elastoplastic compressibility and the elastic compressibility. Figure 3.1.7 compares the recovery rates obtained for the various simulations. We observe widely differing changes between the conventional reservoir simulations and the coupled geomechanical-reservoir model. The latter calculation predicts far higher recovery than the reservoir calculations at start of simulation (which indicates that the reservoir calculation underestimated the effects of pore compression), then lower recovery after a long period of time (which indicates that the reservoir calculation overestimated the effects of pore compression).

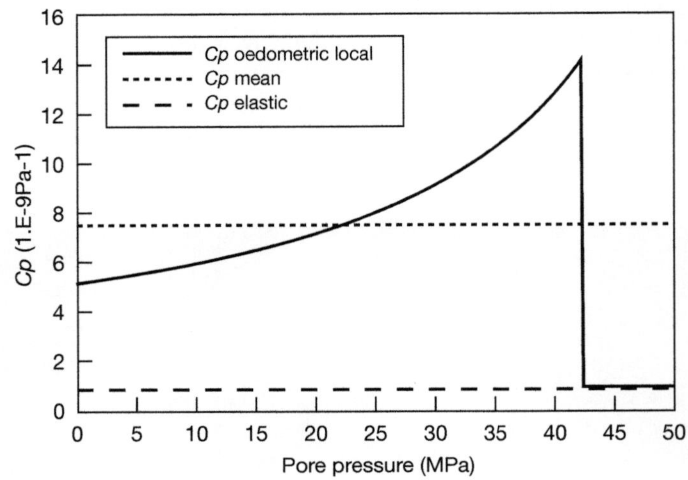

Figure 3.1.6

Various pore compressibilities used in conventional reservoir calculations.

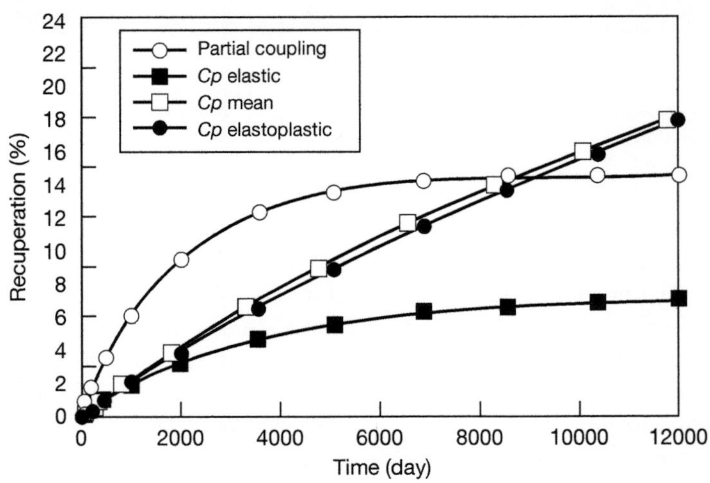

Figure 3.1.7

Comparison of the coupled calculation with various conventional reservoir calculations using various compressibilities [Bévillon, 2000].

3.1.2.6 Example of coupling in permeability

This section illustrates the explicit loose coupling methodology with the reservoir model permeability updated using the geomechanical model of a faulted reservoir. The results presented were obtained using the ATHOS^{TM} reservoir simulator and the VISAGE^{TM} geomechanical simulator [Longuemare *et al.*, 2002]. The reservoir studied is a highly heterogeneous and compartmentalised limestone reservoir located in the Middle East. The geomechanical study was conducted to find out whether the changes in effective stresses and temperature resulting from production and injection of cold water could explain premature watercuts.

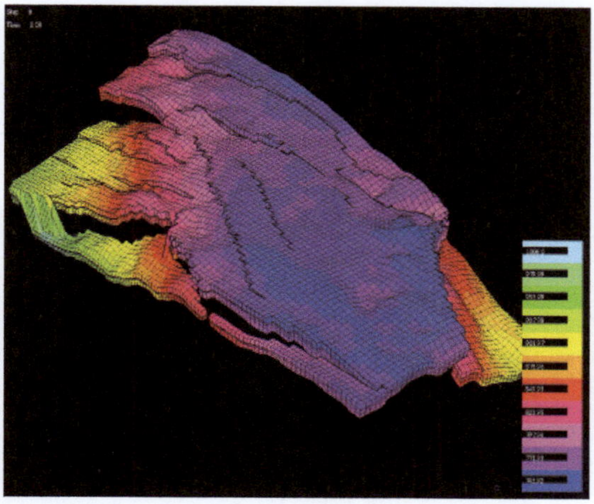

Figure 3.1.8

Reservoir geometry [Longuemare *et al.*, 2002].

Figure 3.1.8 shows the reservoir geometry taken into account in the reservoir simulator. The reservoir model is a simple porosity model including faults whose permeability is described by transmissivity multipliers. The faults taken into account in the geomechanical model associated with the reservoir model are shown on Figure 3.1.9. Figure 3.1.10 shows the distribution between the two types of rock identified in the reservoir according to their dolomite content. The first type, with low dolomite content, has smaller drained Young's modulus and effective cohesion (E_o = 12 GPa, C' = 4 MPa) than the second (E_o = 24 GPa, C' = 11.3 MPa).

The reservoir simulations model the thermohydraulic behaviour of the reservoir over a period of 5 years. The geomechanical model calculates the associated stress variations and the strains normal and tangential to the fault planes (Figure 3.1.11). The latter are related to the permeability changes in the faults via a phenomenological law [Koutsabeloulis and Hope, 1998]. The permeability changes in the faults are then used to update the fault transmissivities included in the reservoir model. Figures 3.1.12 and 3.1.13 show the fault transmissivity multipliers of the reservoir model after 5 years production. The activated faults (whose permeability is increasing) correspond to the faults on Figure 3.1.11 stressed due to the pressure and temperature variations in the reservoir and their orientation with respect to the direction of maximum stress.

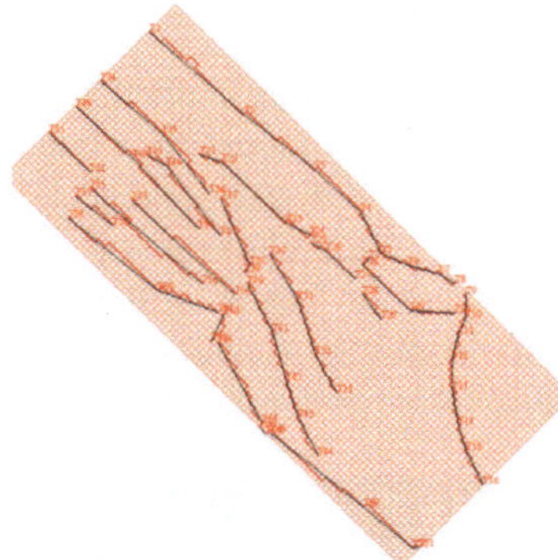

Figure 3.1.9

Faults of the geomechanical model [Longuemare *et al.*, 2002].

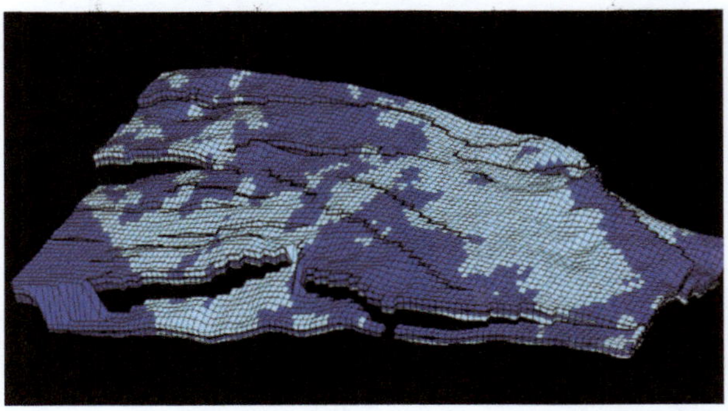

Figure 3.1.10

Distribution of the two types of rock in the reservoir (dark blue E = 12 GPa, light blue E = 24 GPa) [Longuemare *et al.*, 2002].

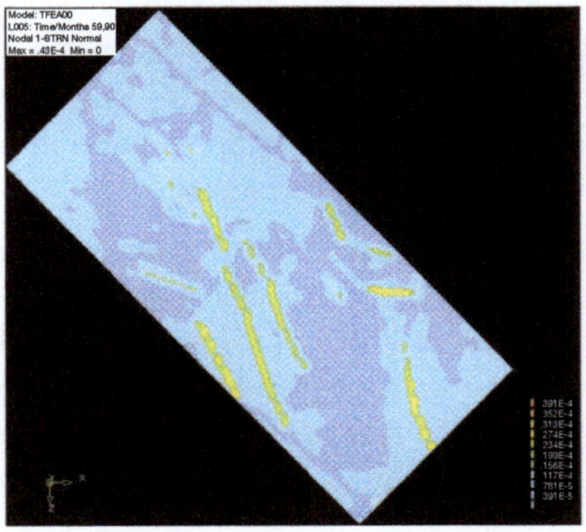

Figure 3.1.11

Strain isovalues normal to the fault planes after 5 years production [Longuemare *et al.*, 2002].

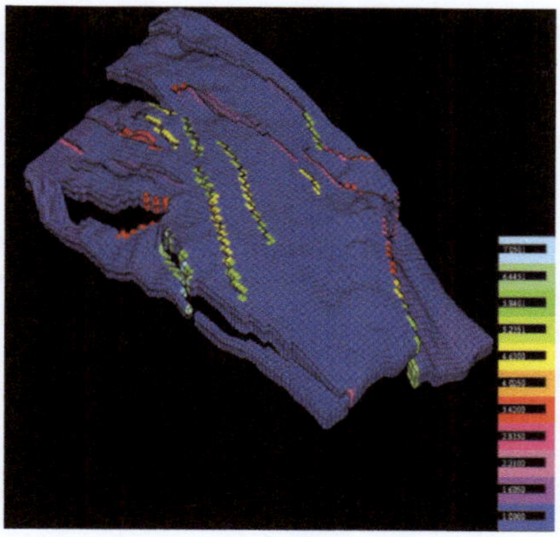

Figure 3.1.12

Transmissivity multiplier in the x direction [Longuemare *et al.*, 2002].

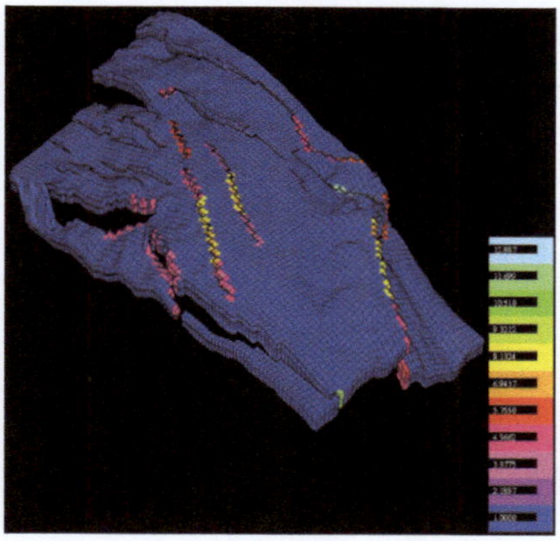

Figure 3.1.13

Transmissivity multiplier in the y direction [Longuemare *et al.*, 2002].

3.1.3 Stress path followed by the reservoir during depletion

Primary production of hydrocarbons leads to a decrease in reservoir pressure. Depletion increases the effective stresses, which finally generates strains inducing modifications of the mechanical parameters and petrophysical characteristics of the rock (note: in this chapter, the stresses are taken to be positive in the direction of compression).

The *in situ* stress changes obviously take place in all directions. Since the minimum horizontal stress can be measured (see Chapter 1.4), the response of a reservoir during depletion is generally described by the ratio between the variations in minimum horizontal stress σ_h and reservoir pressure p_p, called the stress path [Addis *et al.*, 1997, 1998] [Kenter *et al.*, 1998] [Hettema *et al.*, 1998] [Santarelli *et al.*, 1998]:

$$\gamma = \frac{\Delta\sigma_h}{\Delta p_p} \tag{3.1.20}$$

with $\Delta p_p < 0$ for depletion. Note that expression (3.1.20) is not based on any behaviour or boundary condition assumption.

Some authors adopt a different definition of the stress path, expressed as a function of the variations in minimum effective horizontal stress σ_h' and effective vertical stress σ_v' [Rhett and Teufel 1992a] [Schutjens and de Ruig 1996] [Hettema *et al.*, 1998] [Schutjens *et al.*, 1998] [Yale and Crawford, 1998]. Giving, assuming elastic behaviour,

$$K = \frac{\Delta\sigma'_h}{\Delta\sigma'_v} = \frac{\Delta(\sigma_h - bp_p)}{\Delta(\sigma_v - bp_p)} \tag{3.1.21}$$

where b is Biot's coefficient.

If we assume that the axial stress does not vary during depletion (no stress arching), we obtain the following relation:

$$\gamma = b(1-K) \tag{3.1.22}$$

3.1.3.1 *In situ* measurements and assumption of uniaxial compaction

The assumption most frequently accepted to estimate the stress path of a reservoir during depletion based on the reservoir rock characteristics is uniaxial reservoir compaction, i.e. with no horizontal strain. If in addition we adopt an isotropic poroelastic behaviour model for the reservoir rock, we obtain:

$$\gamma = b\frac{1-2\nu_o}{1-\nu_o} \tag{3.1.23a}$$

$$K = \frac{\nu_o}{1-\nu_o} \tag{3.1.23b}$$

where ν_o is the drained Poisson's ratio.

In situ measurements taken on different fields (sandstones and chalks) show that the stress path $\gamma = \frac{\Delta\sigma_h}{\Delta p_p}$ followed by a reservoir during depletion varies between 0.2 and 1 [Addis *et al.,* 1998] and may be quite different from the path predicted assuming uniaxial compaction. Thus, for the Groningen field (consolidated sandstone with 15%-20% porosity), the stress path estimated using *in situ* measurements is $\gamma = 0.4$, to be compared with the value of 0.8 obtained using equation (3.1.23a) [Hettema *et al.*, 1998]. The compilation of 20 years *in situ* data gives a stress path $K = 0.2$ for the Ekofisk field, whereas uniaxial compaction tests in laboratory lead to values ranging from $K = 0.4$ for chalks in the Tor formation, to $K = 0.5-0.6$ for the Lower Ekofisk formation (Rhett and Teufel, 1992a).

It is important to note that use of expressions (3.1.23) to predict the stress path during depletion from laboratory measurements is based on strong assumptions:

- uniaxial reservoir compaction,
- elastic behaviour of the rock during depletion,
- homogeneous drained Poisson's ratio at reservoir scale, identical to that measured at laboratory scale.

Various factors influencing reservoir behaviour during depletion question the above assumptions; they are described in the next paragraph. From the point of view of scale effects between the laboratory samples and the reservoir, note for example that in order to calibrate the predicted subsidence values with *in situ* data, the rock stiffness obtained in the laboratory must be divided by 2 for Norwegian chalk reservoirs and, inversely, multiplied by 2 for the Groningen field sandstone [Santarelli *et al.*, 1998]. Upscaling from laboratory data to reservoir scale is a central problem, and one of increasing interest to rock mechanics engineers in the petroleum community.

3.1.3.2 Factors influencing reservoir behaviour during depletion

Influence of surrounding formations and reservoir geometry

Analytical and numerical models taking into account the reservoir and its surrounding formations (upper, lower and side layers) have demonstrated that the rocks surrounding the reservoir could have a significant influence on its behaviour during depletion. Refer to [Bévillon, 2000] for a numerical approach combining a reservoir model, describing the flow of the various fluid phases, and a geomechanical model, providing a rigorous mechanical framework to describe compaction.

For a reservoir represented by an ellipsoidal inclusion, Addis *et al.* (1998) calculated that the stress path γ decreases when the ratio between the drained Poisson's ratios of the reservoir rock and the host rock increases (all other data remaining constant). For a disc-shaped reservoir of radius R_r, included in a host rock with a lateral condition of zero horizontal strains at radius R_h in the host rock, they obtain:

- for $\frac{R_r}{R_h} = 0.1$, a stress path $\gamma = 0.84$ independent of the ratio between the drained Young's moduli of the reservoir rock and the host rock;
- for $\frac{R_r}{R_h} = 0.01$, a stress path varying between $\gamma = 0.77$ for $E_0^r = E_0^h$ and $\gamma = 0.93$ for $E_0^r = \frac{E_0^h}{20}$.

Kenter *et al.* (1998) generated a 3D mesh of the Shearwater field using seismic data, producing a better representation of the true geometry of the reservoir, the surrounding layers, and the fractures. They demonstrated that, due to the special geometry of the reservoir studied (inclined and elongated), depletion induces stress arching which reduces the stresses at the top of the reservoir and increases them at its boundaries.

Plastic behaviour of the reservoir rock

In case of reservoir repressurisation, the stress path may be quite different from the path followed during depletion, thereby indicating stress path irreversibility. This is the case in particular of the North Sea poorly consolidated sand reservoir studied by Santarelli *et al.* (1998): they obtain $\gamma_{depletion} = 0.4$ or 0.7, depending on the zone considered and $\gamma_{represurisation} \approx 0$ (i.e. $\Delta\sigma_h \approx 0$). The Venice bay follows a stress path during repressurisation one order of magnitude lower than the stress path during depletion.

Kenter *et al.* (1998) represented the behaviour of the Shearwater field using a Cam-Clay model. The simulations carried out demonstrate that the curve showing the change in minimum horizontal stress as a function of the pressure decrease changes slope when the reservoir rock elastic limit is reached. The stress path (calculated according to the stress and pressure state before depletion) changes from $\gamma = 0.7 - 0.8$ when the plasticity criterion is reached to $\gamma = 0.3 - 0.4$ at end of depletion.

Influence of fractures

The presence of fractures may result in a non-homogeneous stress path in the reservoir. Santarelli *et al.* (1998) report, in particular, measurements taken on a North Sea poorly consolidated sand reservoir which reveal the existence of two distinct zones: the first zone follows a stress path $\gamma \approx 0.7$, while the second follows a lower stress path $\gamma \approx 0.4$ due to the presence of a greater number of fractures (faults observable on the seismic data).

3.2 ENVIRONMENTAL ASPECTS

3.2.1 Subsidence

3.2.1.1 Introduction and famous examples

Subsidence is the lowering of the ground surface due to natural causes or induced by human activity.

Natural subsidence may be caused by lowering of the lithosphere. In this case, there are three types of associated subsidence:

- tectonic subsidence, which may be related to crustal thinning (mainly tensile stress regime) and synclinal folding (mainly compressive stress regime) [Perrodon, A., 1989]. This type of subsidence is encountered in rifts and active margins;
- thermal subsidence, resulting from thermal contraction of the lithosphere;
- flexural subsidence, related to a topographic load acting on the lithosphere (e.g. presence of relief) and generating local bending of the lithosphere.

In addition to lowering of the lithosphere, other natural phenomena may contribute to natural subsidence. This is the case in particular of natural compaction of sediments resulting from their consolidation with expulsion of the interstitial water. This compaction, which maybe particularly pronounced in case of coals and clays, may generate subsidence of the ground surface. Dissolution of underground formations may also contribute to the appearance of subsidence. For example, dissolution or carbonate rocks in contact with rainwater loaded with carbonic acid or dissolution of a salt layer by an unsaturated brine.

Natural subsidence nevertheless remains a slow phenomenon, the annual rate of subsidence generally varying from a fraction of a millimetre to a few millimetres per annum.

Subsidence related to human activity is generally faster. It may be caused by mining (extraction of coal, salt or other minerals) or generally by the production of *in situ* fluids. In the latter case, we can distinguish between exploitation of aquifers and production of hydrocarbons from a reservoir. Generally speaking, measurement of anthropogenic subsidence includes natural subsidence. This chapter focuses on subsidence caused by the production of underground fluids. The articles by Christensen *et al.* (1988) and Boutéca *et al.* (1996) report the main cases of subsidence related to pumping in aquifers and production of hydrocarbons. Some of the most famous cases of subsidence will be mentioned below.

A. Subsidence in the city of Venice, Italy

The city of Venice is slowly sinking with respect to sea level, a problem due to three phenomena: the rise in the average level of the Adriatic sea (eustatism), natural subsidence related to consolidation of deposits in the lagoon substrate, and subsidence induced for exploitation of the aquifer for industrial purposes. The amplitudes of these phenomena differ considerably [Gatto and Carbognin, 1981]. From 1908 until 1980, the average sea level rose by 9 cm while during the same period, natural subsidence was about 3 cm. Exploitation of the aquifer started in 1925, resulting in maximum anthropogenic subsidence of about 14 cm in 1969. Pumping ceased in 1969, arresting subsidence and causing elastic rebound of the ground surface of about 2 cm in 1975. Due to these three phenomena, the sea level rose by about 23 cm, a highly critical situation for Venice whose average altitude is about 1 metre above sea level [Lewis and Schrefler, 1998].

B. Subsidence of San Joaquin valley, California, United States

Subsidence of the land surface in San Joaquin valley, California (USA), is mainly due to intensive pumping of the aquifer to irrigate farmland. This subsidence is worsened by subsidence related to the compaction of surface sediments which had little exposure to water after deposition, and which compact in contact with irrigation water. This subsidence varies between 1.5 to 3 metres and may locally exceed a value of 4.5 metres [Bull, 1964; Lofgren, 1975]. Subsidence due to the production of oil and gas reservoirs (less than 1 metre) and tectonic subsidence contribute to a lesser extent to lowering the ground surface. Subsidence due to extraction of groundwater started in the 1920's, reaching a maximum level of 9.1 metres in 1981 (Figure 3.2.1). In 1970, 13,500 km^2 of the San Joaquin valley were affected by subsidence, exceeding 0.3 m over an area of approximately 11,000 km^2 [Poland and Lofgren, 1984]. This subsidence is mainly due to compaction of highly compressible clay beds, resulting from the increase in effective stress associated with the pressure drop in the aquifer. Although the ground surface subsided slowly, there are numerous negative effects: fractured well casings, damage to roads and canals and irrigation networks due to differential subsidence associated with high horizontal strains.

Figure 3.2.1

Joseph Poland beside a telephone pole in San Joaquin valley approximately marking the position of maximum subsidence in 1981 (Photo: Galloway *et al.*, 1999, p. 23).

C. Subsidence of the Bachaquero, Lagunillas and Tia Juna fields in Venezuela

These three heavy oil fields are located on the east coast of Lake Maracaibo, still known as the Bolivar coast. Production of the Lagunillas field started in 1926 and the first signs of subsidence, flooding due a breech in the dyke, were observed in 1929. Subsidence above the three fields affects an area of about 450 km^2; maximum subsidence of 4 to 5 metres in 1986 was measured above each field. Subsidence above the Lagunillas field is the most problematic since there is significant subsidence near the coastal dyke [Finol and Z. A. Sancevic, 1995]. The dyke has had to be raised on several occasions and a drainage and pumping network has been installed to evacuate the rainwater which may build up in the subsidence bowl.

D. Subsidence of the Wilmington field, California, USA

The Wilmington reservoir is an oil field consisting of seven producing zones divided by 5 main faults. Intensive production of the field started in 1941 and 1942, leading to maximum subsidence of 9 metres in 1968 and affecting a total land area of 75 km^2. Water injection started in 1958, reducing the rate of subsidence from 75 cm/year in 1952 to 0 cm/year in 1968. Subsidence generates large horizontal displacements. On the Wilmington field, a horizontal displacement of 3.66 metres was measured between 1937 and 1966. The consequences of ground subsidence were extremely expensive for the town of Long Beach, estimated at 100 million dollars in 1962 [Marsden and Davis, 1967]. In particular, subsidence caused major damage to the roads and buildings in the town, to the production wells, and flooding. Dykes and retention walls had to be built and structures raised. The subsidence also created earth tremors of medium amplitude (2.4 to 3.3) due to the sudden release of shear stresses in the soft shaly levels.

E. Subsidence of the Groningen field, The Netherlands

Groningen, a vast 900 km^2 gas field in The Netherlands, went into production in 1963. The subsidence area is an ellipsoid of 56×70 km^2; subsidence measured about 8 cm in 1982 for a depletion of about 14 MPa [Hettema *et al.*, 2002]. Subsidence of the Groningen field is difficult to manage due to the low ground level and the proximity of the sea, the maximum possible subsidence being 25 cm to 30 cm. Greater subsidence would involve digging a very expensive draining channel. The Groningen subsidence is a striking example of how important it is to monitor subsidence measurements and predict long-term subsidence. An extensive measuring system has been set up to monitor the compaction and subsidence. The subsidence monitoring system is based in particular on 8 annual levelling measurements over a large area and 14 wells wired down to a depth of 400 metres measuring the compaction of Tertiary and Quaternary sediments. Prediction of the subsidence above the field proved difficult. The first estimation which predicted 90 cm subsidence in 2000 soon turned out to be incorrect, since the measurements were three times less than the predictions [Geertsma, 1989]. The prediction models were later improved to take into account the effects of loading rate on the compaction of sandstones and their degree of cementation [De Waal and Smith 1988, Van Hasselt 1992]. The difference between prediction and measurement was resorbed by these improvements and subsidence at end of operation was reestimated at 60-70 cm.

F. Subsidence of the Ekofisk field, North Sea, Norway.

The Ekofisk reservoir includes the Ekofisk (Danian) and Tor (Cretaceous) formations, composed of highly porous chalk (frequently above 40%) and separated by a low-porosity layer (tight zone). The high porosity of the chalk at the extreme depths of the formations can be explained by the abnormally high pressure in the reservoir. The Ekofisk and Tor formations exhibit low matrix permeability of the order of the millidarcy, and the productivity of the formations is due to a network of natural fractures. The reservoir overburden is heterogeneous, but consists mainly of clay and shales. Production from the oil field, which started in 1971, increased significantly from 1974.

The first subsidence problems appeared end 1984, maximum subsidence was about 3 metres at end 1985, with a subsidence rate of 40 to 45 cm/year. Numerous subsidence-induced casing failures were observed in the reservoir and the overburden [Yudovich and Chin, 1989]. In addition, due to the marine environment the subsidence consequences were more expensive. During the summer of 1987, seven platforms had to be raised 6 metres due to the significant decrease in the air gap between the platform and the sea [Sulak and Danielsen, 1989]. Seabottom subsidence results from reservoir compaction associated with elastic then plastic strains (pore collapse). This strain generates a high irreversible reduction of the porosity when the pore pressure decreases [Johnson *et al.* 1989]. Reservoir subsidence may be increased by consolidation of clays in the overburden. Several reinjection operations were carried out to limit the subsidence. Some of the gas produced was reinjected from 1985 to 1987 [Sulak and Danielsen, 1989]. While this limited the subsidence rate to about 26 cm/year, it had to be stopped for financial considerations.

Water injection was then carried out to stabilise the pressure in the reservoir and enhance hydrocarbon recovery. Water injection was first started in the Tor formation from end 1987 then extended to the Ekofisk formation from 1994. In 1996, the reservoir pressure was perfectly stabilised (even very slightly increased) although reservoir subsidence continued, reaching a maximum value of about 7 metres (Figure 3.2.2). This permanent strain of the chalk at constant pressure was put down to water weakening (weakening caused by the increase in water saturation of the chalk) [Maury *et al.*, 1996; Sylte *et al.*, 1999; Matà, 2001].

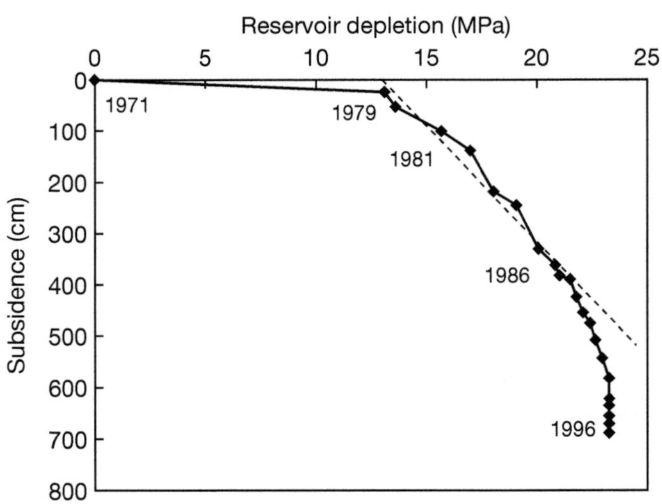

Figure 3.2.2

Subsidence as a function of depletion for the Ekofisk field (Hetema *et al.* 2002).

3.2.1.2 Analysis of subsidence related to the production of underground fluids

Production of underground fluids is generally accompanied by a significant decrease in the pressure of the fluids saturating the reservoir rock pores. This decrease in pore pressure increases the effective stress imposed on the porous medium. This leads to settling or compaction of the reservoir rock, which may be quite considerable in case of highly porous, brittle or poorly cemented, highly compressible rocks and plastically deformable rocks. The parameters influencing reservoir compaction are the pressure drop in the reservoir, its thickness and mechanical behaviour.

While reservoir compaction is a determining factor in the subsidence process, it is not the only one. If, for example, the reservoir is covered by an overburden which is rigid and therefore difficult to deform, the displacements of the reservoir layers resulting from reservoir compaction are totally absorbed by the overburden and no subsidence is generated at the surface. Subsidence only appears if the geometric characteristics (especially thickness) and mechanical characteristics of the overburden allow transmission of the reservoir displacements up to the land surface [Christensen *et al.*, 1988].

Large, shallow reservoirs covered by an overburden of very low rigidity are therefore most prone to subsidence. Since compaction is generally irreversible, this is also true of subsidence. If no other phenomena are involved, maintaining pressure in the reservoir will stop compaction and subsidence. Generally, however, repressurising the reservoir rock will only reduce the subsidence very slightly (elastic rebound) since an increase in reservoir pressure is accompanied purely by elastic strains (Settari, 2002, for example).

Figure 3.2.3 is a simplified representation of the horizontal (U_h) and vertical (U_v) displacements associated with subsidence. The generally dish-shaped subsidence area is called a subsidence bowl. The vertical displacement is zero at large distances from the subsidence zone, progressively increasing up to is maximum value at the centre of the subsidence zone. The horizontal displacement is directed towards the point of maximum subsidence. The maximum horizontal displacement is located about halfway up the flanks of the subsidence bowl, from where it decreases progressively to zero at both the centre and periphery of the bowl.

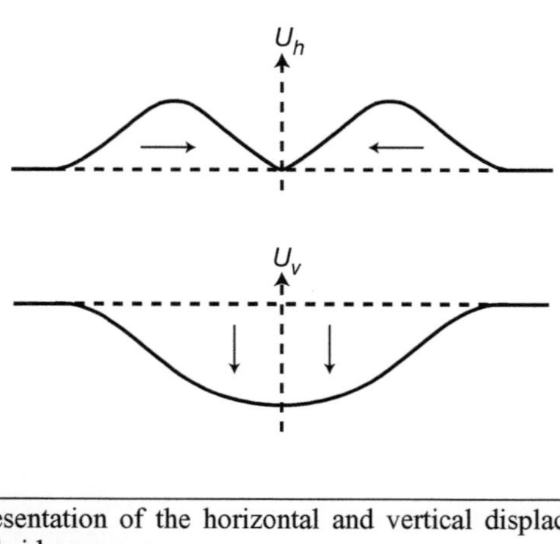

Figure 3.2.3

Idealised representation of the horizontal and vertical displacements on the surface of a subsidence zone.

A delay of a few months to a few years is frequently observed between the pressure drop in the reservoir and its consequences on the surface in terms of subsidence. As shown on Figure 3.2.4, the linear poroelastic model cannot account for this delay, which must therefore be attributed to nonlinearity. This delayed effect can be explained by the usual viscoelastic behaviour of the rocks or by compaction of possible clay levels in the reservoir.

Due to the delay between reservoir depletion and the start of subsidence, long-term monitoring of the displacements will be required after production stoppage and a subsidence measuring system will therefore have to be maintained for many years.

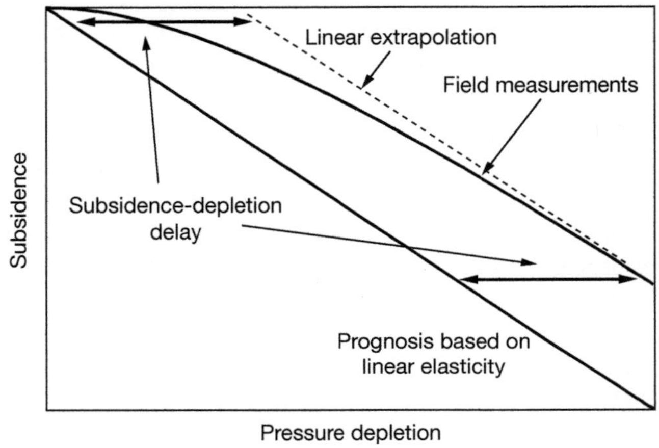

Figure 3.2.4

Diagram of the subsidence-depletion delay effect [Hetema *et al.,* 2002].

The main risks associated with subsidence are illustrated in the above-mentioned examples. Subsidence is a serious problem for offshore platforms (reduced air gap) and near seas and lakes (risk of flooding). In addition, Boutéca *et al.* (1996) point out that the risks are highly dependent on the environment: ten centimetres subsidence is critical at Groningen, for example, but quite acceptable at other fields. Vertical subsidence is associated with horizontal displacements which may damage buildings and surface facilities. The UK National Coal Board classifies damage according to the size of the structure and the amplitude of the displacements on the ground [Geertsma and Van Opstal, 1973]. Horizontal and vertical strains between the surface and the reservoir may also fracture casings, destroy numerous boreholes and reactivate faults.

3.2.1.3 Subsidence measurement and prediction

A. Subsidence measurement

Boutéca *et al.* (1996) propose a review of subsidence and compaction measurement methods. On the ground, subsidence is traditionally measured by levelling survey or GPS positioning. More recently, Brink *et al.* (2002) used an InSAR (Interferometric Synthetic

Aperture Radar) method to determine subsidence rates at the Lost Hills field in California. This technique, based on processing images obtained using SAR radar on board satellite, represents a powerful ground movement estimation tool. At sea, measurement can be carried out less precisely by measuring the height variations between the platform and sea level. Measurements of underground displacements in the reservoir and the overburden can also be used. These measurements can be taken using conventional log analyses (measurement of natural gamma radiation) or specialised log analyses (variation in casing lengths, measurement using radioactive marker techniques, etc). For shallow fields, reservoir compaction can be measured using a taut wire at the bottom of the well.

B. Subsidence prediction

Geertsma (1973) proposed an analytical model to provide an estimation of the subsidence resulting from compaction of a cylindrical reservoir, of thickness h and radius R located at depth z. Reservoir compaction is given by:

$$C = \Delta h = \int_0^h c_m(z)\Delta p(z)dz \tag{3.2.1}$$

with c_m the reservoir rock compressibility in oedometric condition (uniaxial condition and constant vertical stress) which can be measured in laboratory and Δp the pressure variation in the reservoir. If c_m and Δp are homogeneous in the reservoir, the reservoir compaction is given by:

$$C = c_m \Delta p h \tag{3.2.2}$$

Geertsma (1973) then gives the analytical expressions of the horizontal and vertical elastic displacements on the surface in case of infinitely rigid underburden, overburden with the same mechanical characteristics as the reservoir and reservoir with uniaxial behaviour. The vertical displacement at the centre of the reservoir can be expressed as follows:

$$S = -2(1-\nu)\left(1 - \frac{\eta}{\sqrt{1+\eta^2}}\right)C \text{ with } \eta = \frac{z}{R} \tag{3.2.3}$$

According to the latter equation, the ratio between maximum subsidence and reservoir compaction (S/C) is determined by the ratio between depth of burial and the lateral extent of the reservoir. Expression (3.2.3) shows in particular that small deeply buried reservoirs (large η) will present no significant subsidences, even if their reservoir compaction is high. On the contrary, large reservoirs at low depths (small η) may be potential candidates for subsidence.

Morita *et al.* (1989) later proposed a semi-analytical solution based on finite element calculations, which does not assume that the reservoir and its confining formations have the same mechanical properties.

The previous solutions are based on simplifying assumptions which do not allow accurate calculation of the subsidence for a given field. In particular, the complex reservoir geometry, heterogeneities in the reservoir and the surrounding formations and elastoplastic constitutive laws must be taken into account to predict subsidence accurately. More accurate subsidence predictions therefore require coupled geomechanical-reservoir simulations (see Chapter 3.1) to precisely describe reservoir compaction during production and the strains in the overburden [Geertsma and Van Opstal, 1973].

Note that this type of modelling requires numerous data not only on the reservoir but also on the surrounding formations (porosity distribution, mechanical properties, etc.) and on the reservoir production. This type of study is therefore reserved for the most serious cases of subsidence (e.g. the Groningen field).

3.2.2 CO_2 sequestration

For the last century or so, human activity, in particular combustion of fossil hydrocarbons, has increased the levels of greenhouse gases in the atmosphere. The CO_2 concentration has increased and could reach 1,000 ppm by the end of the century if no steps are taken to reduce emissions. Experts agree on the fact that this increase is responsible for noticeable climate changes and variability, with potentially disastrous impacts on the ecosystems and human activities.

As pointed out in the 2005 IPCC (Intergovernmental Panel on Climate Change) report, Carbon Capture and Storage (CCS) is an efficient way of reducing CO_2 emissions. If implemented, the rate of greenhouse gas emission into the atmosphere could start to stabilise significantly in the near future.

Storage in end-of-life oil or natural gas reservoirs offers several advantages, the most important being that it is a known technique and virtually leaktight. These natural reservoirs have proven their ability to store hydrocarbons for several millions of years. CO_2 has already been reinjected for many years in depleted reservoirs to reduce the oil viscosity, improve its mobility and enhance recovery. While this is an EOR (Enhanced Oil Recovery) technique, for sequestration of large quantities of CO2, the pressures considered exceed the pressures that existed *in situ* before any human intervention, which is never the case with EOR.

CO_2 storage in deep aquifers represents another possibility. Numerous deep aquifers exist, onshore and offshore, some with areas of up to several thousand km^2. Formed from porous, permeable rocks, filled with brine unsuitable for human consumption, they could store large quantities of CO_2. Some are already used for this purpose (for example the Utsira formation in the North Sea).

3.2.2.1 Short-term geomechanical risks

CO_2 injection will therefore be carried out on a massive scale, under pressure. A distinction must be made between the local pressure increase at the injectors and the average pressure increase in the reservoir. These pressure increases modify the *in situ* stresses locally and globally, thereby deforming the reservoir and the surrounding layers with the following potential consequences:

- hydraulic fracturing,
- reactivation of faults or fractures,
- shear failure,
- possible leaks to the surface.

Hydraulic fracturing may be increased by the fact that CO_2 injection will generate *a priori* thermal perturbation with a cold front around the injectors. In some cases, fracturing may propagate in the cap rock (see Chapters 2.4 and 2.5).

Pressure injection reduces the effective stresses normal to the faults and fractures and therefore reduces their shear strength. The injector wells must be located far away from faults.

Shear stresses may be created at the top of the reservoir at the interface with the cap rock [Hawkes *et al.*, 2004].

3.2.2.2 Modelling

To quantify the risks, the most reliable approach is to conduct precise coupled geomechanical-reservoir simulations. As when calculating subsidence, it will be essential to take into account the geometry and mechanical properties of the reservoir and the surrounding formations.

Knowledge of the initial *in situ* stresses is also fundamental [Hawkes *et al.*, 2005].

The few studies already conducted indicate possible surface uplift in case of shallow reservoir or aquifer (Figure 3.2.5) [Rutqvist *et al.*, 2007, Vidal-Gilbert *et al.*, 2009].

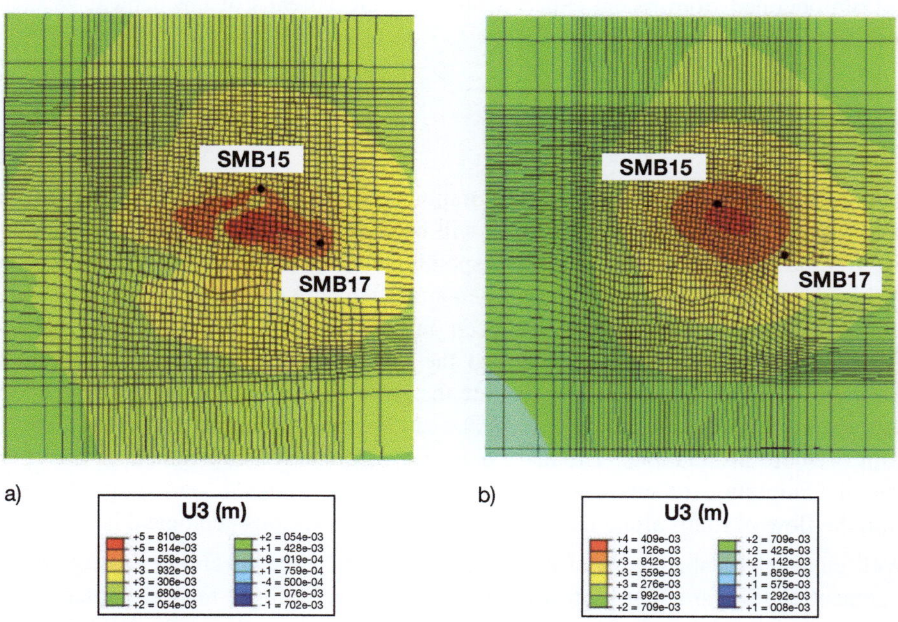

Figure 3.2.5

Variations of vertical displacement (m) a) at the reservoir top and b) at surface during CO_2 injection [From Vidal-Gilbert *et al.*, 2009].

3.2.2.3 Medium-term geomechanical risks

Injection of supercritical CO_2 in a geological reservoir will necessarily modify the physico-chemical equilibrium in the host rock. Dissolution of CO_2 in the formation water causes acidification. This phenomenon accelerates the dissolution-precipitation reactions and may modify the mechanical and hydraulic properties of the rock. The fluid characteristics change on moving away from the injector well and will also change over time, passage of the acid reactive front, then a drying or dehydrating front. We can expect to find dissolution phenomena with more or less fingered profiles as we move further away. The dissolved materials may also precipitate out again at a variable distance from the initial dissolution zone. These phenomena could change the geomechanical properties of the host rock.

In carbonate rocks, with the high injection rates expected around the well, the development of networks of wormholes could be observed [Egerman *et al.*, 2006]; farther away, lower rates would lead to more compact dissolution figures.

In sandstone reservoirs with carbonate cement, CO_2 injection may generate a number of damaging phenomena. In particular, progressive dissolution of a carbonate cement could lead to considerable mechanical weakening of the rock.

Generally, alteration of the reservoir matrix by the CO_2-brine mixture could result in compaction of the reservoir rock and development of cracks in the reservoir and the cap rock. Data obtained from recent experiments show clear trends of mechanical weakening [Bemer, 2008].

3.2.3 Well abandonment

Wells are drilled for various reasons: exploration, oil and gas production, water injection, storage, etc. At some stage, all these wells will be permanently abandoned. Producer wells have a lifetime of about twenty years, corresponding to the production period of an oil field. After this phase, the well must be abandoned and plugged with suitable material, typically cement-based. Even though the well has been plugged, there is a serious risk of leakage in the long term which represents a threat to the environment. For example, migration of hydrocarbons along the borehole may pollute the groundwater, the marine ecosystem or the atmosphere in case of toxic gases. These leaks can be sealed, but repair costs are extremely high for offshore installations. The main objective of oil well abandonment is therefore to prevent the migration of any hydrocarbon reserves remaining at end of production by limiting the flow of fluids along the borehole between the various geological layers.

Well closure and abandonment is therefore a critical issue regarding protection of the environment and in particular protection of water resources. Since the techniques involved in closure require a detailed analysis of the geological context, the well, variations in bottomhole conditions expected over the long term, reservoir fluids, etc., each abandonment procedure is unique.

3.2.3.1 Abandonment procedure

The operator responsible for well closure and abandonment must comply with local laws (aimed at protecting the environment) while also respecting financial constraints. It is not financially viable, for example, for the operator to pour a cement plug down the entire depth of the reservoir and also seal all the annuli from well bottom up to the surface. The abandonment procedure varies depending on whether or not some of the steel tubes (production tubing and some of the casings) are abandoned with the well. Offshore wells are frequently abandoned with some of their steel tubes if no drilling equipment is available.

Kelm and Faul (1999) recommend installing several plugs to optimise well isolation (see Figure 3.2.6). In particular, these authors recommend installing cement plugs at the casing perforations, above the oil zones located above the perforations, at abnormal pressure zones and above the noncemented annuli. Plug length varies from 15 to 60 metres, the longest plugs being recommended for the abnormal pressure zones and the surface plugs (i.e. the last plug preventing surface pollution).

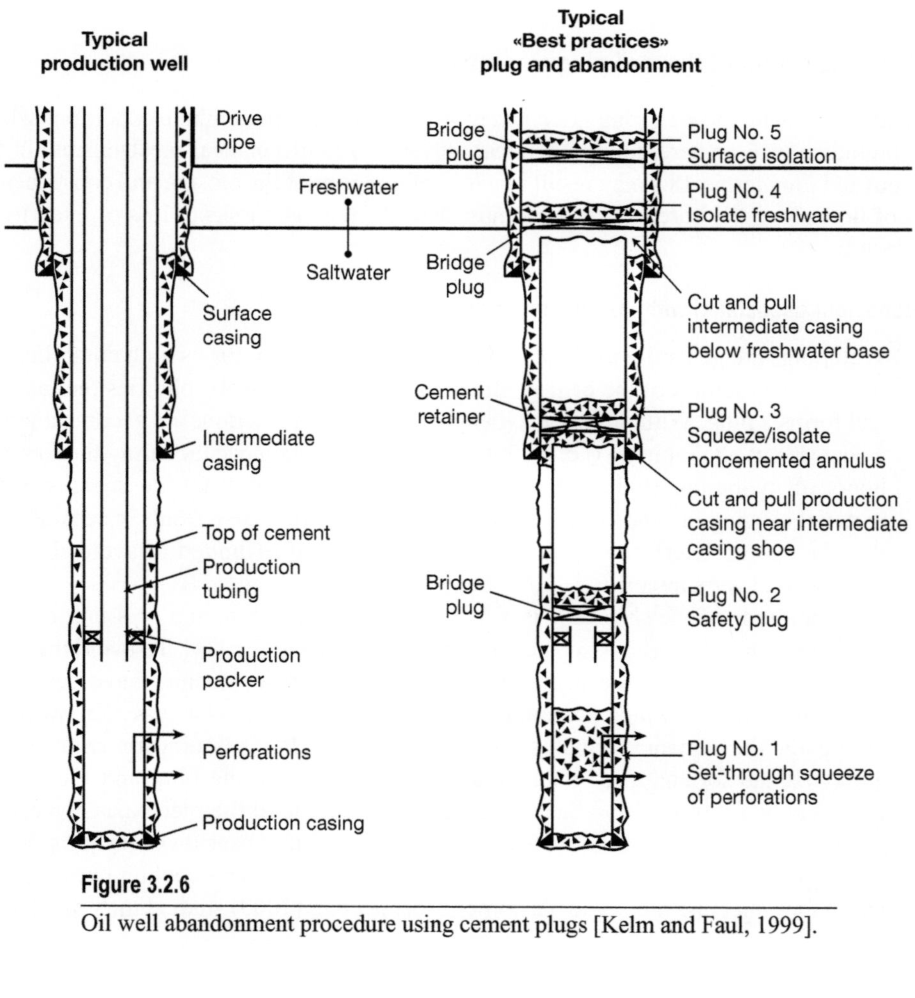

Figure 3.2.6

Oil well abandonment procedure using cement plugs [Kelm and Faul, 1999].

Installation of the well plug is a determining factor in closure quality. Wells are generally plugged with Portland cement to which sand and bentonite (for expanding cements) can be added. Cement set accelerators can also be added, as well as dispersing agents to improve the durability of the materials. For very large diameter wells, operators may also use concrete. With water wells, use of bentonite-based clay materials is sometimes recommended [Calvert and Smith, 1994]. Presence of fluid in the well makes installation of the cement plug difficult since it favours dilution and the creation of preferential flow paths of the cement slurry [Smith *et al.* 1984]. The density difference between the cement slurry and the fluid in the well may cause the cement slurry to flow to the bottom of the well, the result being that no plug is really formed [Calvert and Smith, 1994, Crawshaw and Frigaard, 1999]. This type of flow can be avoided by installing a mechanical plug under the cement plug, to ensure the cement slurry remains in place while setting. The installation procedure must also ensure perfect adherence between plug and wall to avoid risks of leakage at the interfaces. To improve adhesion between the casing and the cement plug, Evans and Carter (1963) and Kelm and Faul (1999) recommend cleaning the tubing before installing the cement plug.

3.2.3.2 Behaviour of the well plug after abandonment

After abandonment, the well plug is subject to variations in the bottomhole conditions which may prejudice well closure due to chemical, physical and mechanical mechanisms. In the medium and long terms, this may result in a loss of integrity of the closure and consequently risks of flow along the borehole. We will now describe the main risks of loss of integrity of the closure.

A. Behaviour associated with installation of the plug

- Well plugs are generally made of cement, a material chosen for its high durability and low cost. Hydration of the cement slurry injected into the well solidifies the material and forms a plug. With ordinary cements, the hydration reactions are accompanied by a decrease in plug volume (i.e. hydration shrinkage) which occurs during the first few hours of hydration. Hydration shrinkage may generate high tensile stresses in the material causing the cement to crack. Plug shrinkage may also result in separation at the plug/casing interface. These two phenomena must be limited or avoided to preserve well closure integrity.
- The amplitude of the shrinkage is affected by the temperature and pressure to which the cementation material is subjected [Parcevaux and Sault, 1984]. To avoid the negative effects of cement shrinkage during setting, service companies have developed expanding cements which prevent the formation of microannular spaces between the casing and the cemented annulus [Seidel and Greene, 1985; Baumgarte *et al.,* 1999]. To avoid the formation of microannular spaces between the plug and the casing, Bosma *et al.* (2000) propose installing an elastomer plug of thermal expansion coefficient greater than that of the surrounding rock. Thermal expansion of the plug occurs due to progressive heating of the rock and the elastomer after a cooling phase carried out before injecting the liquid elastomer. Bosma *et al.* (2000) studied in particular the

influence of this cooling phase to avoid excessive thermal expansion of the plug, which could damage the primary cementation of the annulus (Figure 3.2.7).

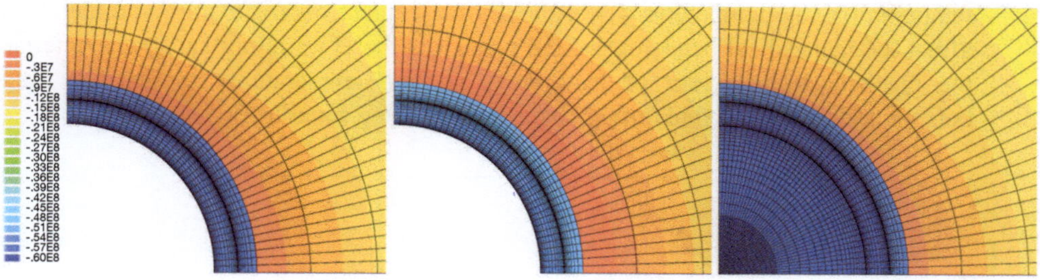

Figure 3.2.7

Evolution of the radial stress around the well a) before cooling, b) after cooling and c) after installing the plug and heating [Bosma *et al.*, 2000].

B. Behaviour associated with reservoir production and abandonment

- Reservoir production modifies the pressure, temperature and stress in the reservoir and more especially around the boreholes. If a well is abandoned and reservoir production maintained, these variations will generate thermal, hydraulic and mechanical loads on the well plug. If production is stopped completely and all wells in the reservoir abandoned, the return to thermal, hydraulic and mechanical equilibrium will be accompanied by temperature, pressure and stress variations which will also act as loads on the plugs. These loads are difficult to estimate since they depend on each individual case. In particular, presence of an active aquifer under the reservoir may cause the pressure to rise in the reservoir; in wells operating with injection of cold water during production, the temperature around the injector well will rise when the well is abandoned and, inversely, in wells implementing enhanced recovery processes based on steam injection, the area around the well will cool down when the well is abandoned.
- The mechanical properties of the well plug must therefore be adapted to variations in bottomhole conditions after abandonment. These variations can be estimated using coupled geomechanical-reservoir simulations which supply the pressure, temperature and stress variations after abandonment of the well and/or the reservoir.
- Using this technique, Mainguy *et al.* (2004) were able to estimate the stress variations in the cap rock of a reservoir where it is recommended to install a plug [UKOOA, 2001]. This study shows in particular that the cap rock behaviour is a determining factor for durability of the well plug, since a reduction in the effective compression stress during the abandonment phase in this cap rock (Figure 3.2.8) induces tensile stress in the plug if it was not initially prestressed. This type of load is particularly detrimental for cement based plugs which have low tensile strength and may also favour separation of the plug from the casing or surrounding rock.
- The behaviour of cement plugs is generally modelled using a constitutive law which combines a smeared cracking model [De Borst and Nauta, 1985] with an elastoplastic

constitutive law in compression. In particular, this is the model used by Bosma *et al.* (1999) and Ravi *et al.* (2002) to study the behaviour of the annuli cemented during the well completion and production phases. In case of deep wells exposed to high temperatures, the temperatures will modify the cement microstructure and therefore affect the mechanical behaviour of the plug. It is therefore important to characterise the mechanical behaviour of well plugs at high temperatures so that the plug behaviour can be modelled during abandonment using suitable models.

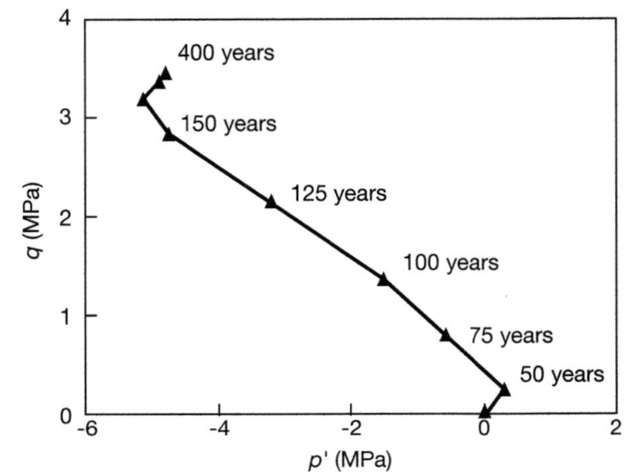

Figure 3.2.8

Example of tensile load applied in the shale caprock above an injector well [Mainguy *et al.*, 2007]. Evolution of the deviatoric and effective mean stresses from end of production to 400 years after abandonment.

C. Plug ageing under high pressure and temperature conditions

– Well plugs installed in the reservoir are exposed to high pressure and temperature conditions which could damage the cement plug in the long term. Ageing of the cement slurry under severe pressure and temperature conditions irreversibly damages its mechanical properties (Young's modulus, breaking strength, etc.). Studies on ageing of cement materials under high pressure and temperature conditions are generally conducted over periods of 1 to 2 years [Eilers and Root, 1974, Noik and Rivereau, 1999]. These data must nevertheless be extrapolated over longer time scales for use in studies on abandonment over longer periods.

D. Leaching of cement plugs

– In contact with a neutral or slightly basic solution (as compared with the highly basic interstitial solution of a cement slurry), the hydrated phases of the cement, in particular portlandite ($Ca(OH)_2$), dissolve. This cement degradation process (called leaching) combines dissolution of the chemical compounds in the cement and transfer of chemical species in liquid phase by diffusion [Adenot, 1992]. Chemical degradation of the cement is accompanied by degradation of its elastic mechanical [Gérard, 1996] and breaking [Carde, 1996] properties. Nevertheless, the aggressive water near the plug must be renewed if degradation of the cement plugs by leaching is to take place. This scenario is only possible with plugs installed near an aquifer or in contact with runoff water.

E. Carbonation of cement plugs

- Cement carbonation designates all chemical reactions between carbon dioxide and the compounds in the hardened cement produced by cement hydration reactions. Carbon dioxide penetrates the cement in gaseous form or in solution. Carbonation occurs in aqueous medium, by reaction of the carbon dioxide dissolved in the interstitial solution of the cement with portlandite that has gone into solution. This reaction first produces calcium carbonate, then finally calcite. C-S-H compounds (silicates) and aluminates may also react with carbon dioxide. Progression of carbonation towards the centre of a cement paste is generally assumed to be proportional to the square root of the carbon dioxide exposure time.
- The mechanical properties of the cement paste deteriorate with carbonation. Chi *et al.* (2002) studied the influence of the depth of carbonate cement on the loss of unconfined compressive strength. Beaudoin (2000) studied the reduction in Young's modulus of a carbonate cement slurry as a function of the increase in its porosity resulting from the dissolution of hydrates. Carbonation of cement plugs installed in reservoirs seems unlikely, however. This type of phenomenon could be observed in cement plugs used to close CO_2 storage wells.

3.3 GEOMECHANICAL MONITORING OF RESERVOIRS

Reservoir monitoring is a production technology designed to measure, check and predict the reservoir performance. Monitoring generally includes measurements of temperature, pressure, flow rate and constituents, as well as surface deformation, 4D seismics and microseismicity which no longer concern the well alone but consider the reservoir as a whole with the surrounding formations. These records can be used to map the progression of a heat front in 3D, identify reactivation of a fracture or fault and check the integrity of the cap rock.

3.3.1 Measurement of surface deformations

Volume variations in the reservoir may cause deformations in all the surrounding geological layers as well as surface deformations. This occurred with subsidence, but it may also be the case when the reservoir is shallow and operated using thermal methods.

Tiltmeters can be installed in a fixed network on the surface or moulded in the wells. Modern tiltmeters can measure angles as small as few nanoradians. They are so sensitive that they record the Earth tides generated by the moon. They have been used to record surface deformations during SAGD processes (Figure 3.3.1).

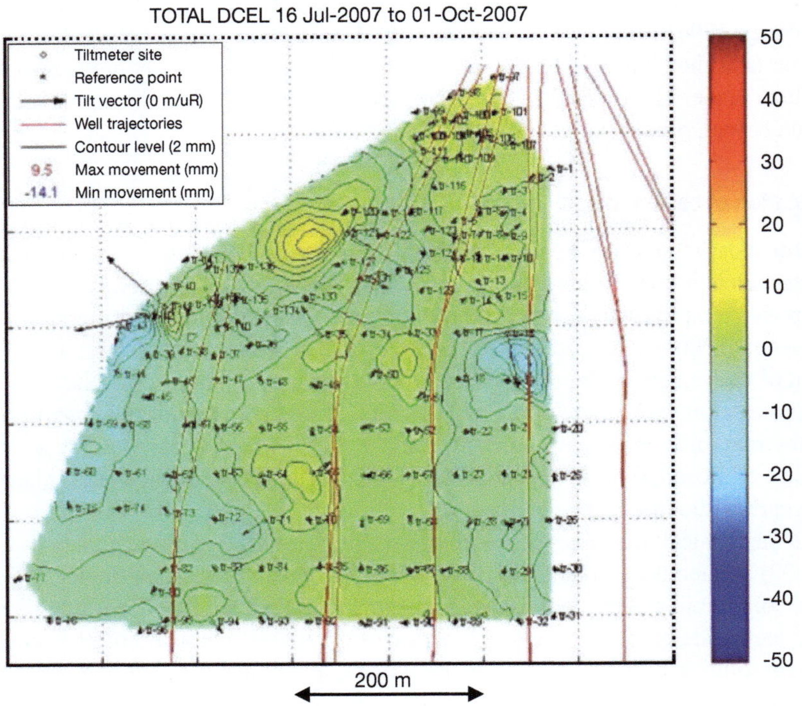

Figure 3.3.1

Surface deformation over a period of 3.5 months as computed from tiltmeter measurements on a SAGD field [Dubucq *et al.*, 2008].

"Precision" GPS instruments and interferometric aperture radar techniques can also be used to measure surface displacements, but with less accuracy.

The InSAR (Interferometric Synthetic Aperture Radar) technique uses two synthetic aperture radars simultaneously, or the same radar used at different times. The point to point phase differences in the images generated are then studied to find the altitude of the land. Centimetric displacements can be measured in areas where the signal remains coherent [Mei and Frosses, 2007].

3.3.2 Seismic monitoring

The seismic technique is the geophysical method most commonly used to define sub-surface structures. The most common implementation of this technique consists in reflection seismic with multiple coverage. This method provides imaging of the sub-surface in 2 or 3 dimensions. Lithology and fluid information can be extracted from seismic data (wave propagation velocities, amplitudes, etc.).

Time-lapse seismic reservoir monitoring is the process of acquiring and analysing multiple seismic surveys, repeated at the same site over calendar time, in order to image fluid-flow effects in a producing reservoir. If each survey is "3-D seismic," then the resulting set of time-lapse data is termed "4-D seismic," where the extra fourth dimension is calendar time [Lumley, 1995].

In addition to 4-D seismic, there are other viable methods of time-lapse seismic monitoring including repeated 2-D surface seismic, surface-to-borehole VSP (vertical seismic profiling), and crosswell tomography.

Figure 3.3.2 shows two seismic cross-sections acquired at different times during the production of the Duri field, Indonesia: before steam injection and after 5 months of steam injection. The last seismic cross-section is the difference between the two previous ones. The well outline visible in the third cross-section is due to a shallow borehole steam leak.The time lapse seismic images show some changes in amplitudes along baseline reflectors, due to a variation in fluid content.

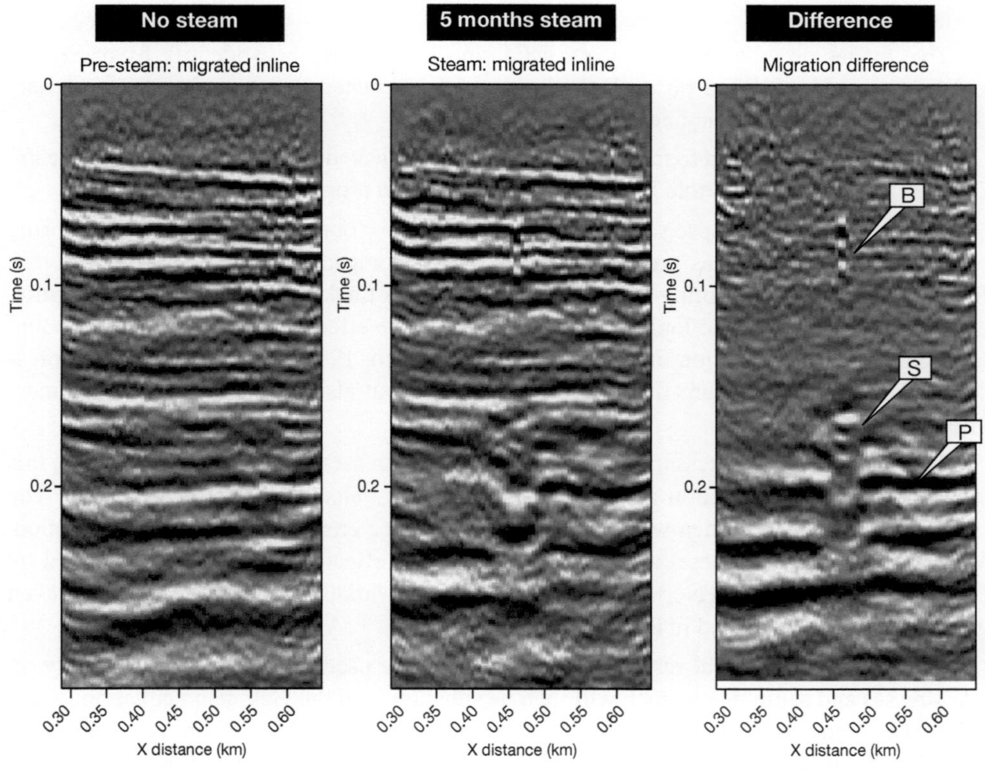

Figure 3.3.2

Seismic cross-sections: (left) before steam injection, (centre) during injection, (right) difference between the two cross-sections. S: steam zone, B: borehole heating, P: polarity reversal [Lumley, 1995].

Similarly, Figure 3.3.3 shows the maps of amplitude differences between the seismic surveys carried out in 2001, 2004 and 2005 on the Christina Lake field operated by SAGD. The amplitude differences indicate the changes in the steam chamber around the wells.

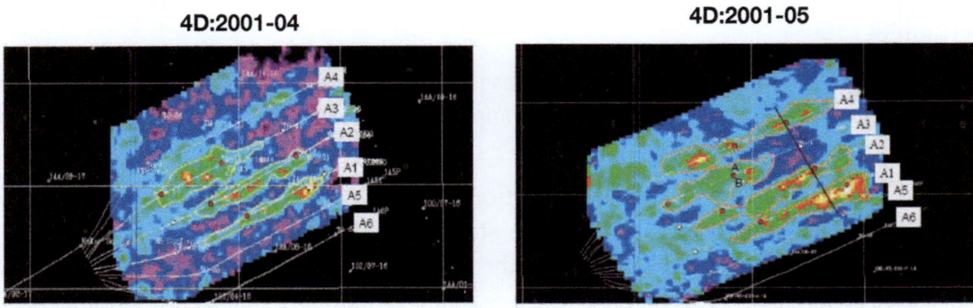

a) 2004 b) 2005

Figure 3.3.3

Maps of amplitude difference a) between 2001 and 2004 and b) between 2001 and 2005 [Zhang *et al.*, 2005].

The images provided by reservoir seismic monitoring may help identify by-passed oil volumes, which will then be drilled. The reserves extracted can therefore be increased, thereby extending the economic life of an oil field.

4D seismic can be used to track the progress of the injected fluid front (water, gas, steam, CO_2, etc.) and may play a role in optimising the injection programs.

More recently, time-lapse seismic has been applied for objectives other than monitoring saturation changes, such as pressure monitoring and compaction detection. Over the past few years, seismic monitoring has been widely used to track fluid movements (saturation effect). This approach is no longer sufficient and pressure effects must be taken into account to improve the interpretation of 4D seismic data [Landro, 1999]. More precisely, the pressure effect does not concern the pore pressure alone, but also the induced geomechanical effects.

Decreasing the pore pressure in the reservoir will increase the mean effective stress; this will lead to rock compaction and increase the P-wave velocity Vp in the reservoir. The overburden and underburden will deform to fill the space created by reservoir compaction. Vp may be reduced in these zones due to the arching effect, a phenomenon observed by Hatchell *et al.* (2004) on a North Sea field. The velocity variations in the cap rock were even greater than those observed in the reservoir.

Coupled geomechanical-reservoir simulations can be used to determine the porosity, *in situ* stresses and saturations which will then be substituted in the Gassmann equation to calculate the P- and S-wave velocity field not only in the reservoir but also in the entire block considered. The studies conducted in this field are highly promising [Vidal *et al.*, 2002, Minkoff *et al.*, 2004].

3.3.3 Passive seismic monitoring

The various terms used, *seismic* or *microseismic monitoring, Passive Seismic Monitoring (PSM)* and *microseismicity* refer to this technique, based on the recording of induced seismic signals. The term induced implies that the seismic activity recorded results from a man-made action or operation on the medium, as opposed to natural seismicity related to tectonics.

Microseismic events correspond to sudden readjustments of mechanical stresses (Figure 3.3.4). Stress rebalancing may be caused by a variation in pore pressure (oil production, storage of fluid in underground aquifer, extraction of material mining, washout), or variations in the temperature of the medium (injection of fluid which is hot or cold with respect to the medium: geothermy, enhanced recovery by water flooding, thermal methods). In some cases, these phenomena may be superimposed.

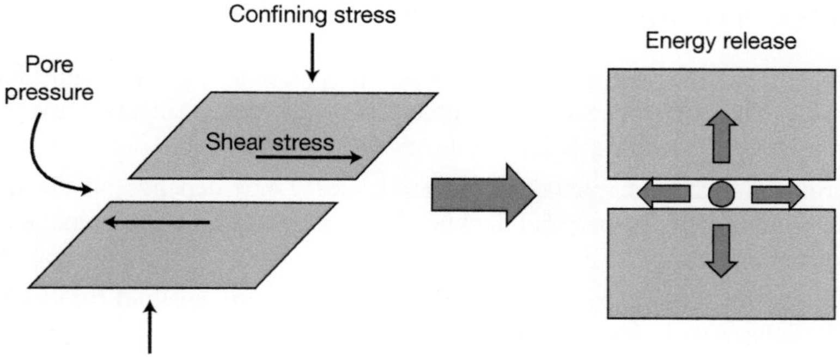

Figure 3.3.4

Microseismic failure mechanism.

Initially observed in Colorado after injection of fluid waste for underground disposal in the 1960's, induced microseismicity soon became a subject of interest to oil producers in order to map hydraulic fractures, the first publications dating back to the 1970's.

In the following years, microearthquakes were observed in the highly depleted oil zones, especially the Groningen (the Netherlands), Gazli (Uzbekistan) and Lacq (France) fields. Since these were naturally seismic risk areas, implementation of monitoring was in some respects expected to distinguish between natural seismicity and induced seismicity. Surface monitoring was carried out, using seismology equipment.

In the 1980's, Passive Seismic Monitoring (PSM) became widely used in the mining industry to prevent the risk of cave-in, especially in South Africa where extraction of various ores (coal as well as coal and diamonds) is carried out at greater and greater depths. In the field of underground storage, this technique has been used to monitor storage in salt cavities.

In these fields, the measurement means are traditional and inexpensive, consisting of geophones connected to recording systems. The frequencies used ranges from a few Hz (less in seismology) to several hundred Hz. The most recent developments take into account three-component sensors.

In the oil industry, the depth of the reservoir may limit the use of this technique due to signal attenuation particularly in the near surface.

Ideally, information derived from interpretation of microseismic data can be used to:

- locate possible sliding over a natural fault plane,
- locate the compacting zones,
- locate the hydraulic fracturing zones and determine the characteristics (width, length, height, orientation, growth, complexity, etc.) of the fractures created,
- identify the natural fractures or faults contributing to fluid displacements,
- map the fluid displacements,
- locate the zones badly drained or not drained by the fluids injected,
- identify the mechanically active levels likely to play a role in well failure (casing),
- map the thermal fronts, etc.

3.3.3.1 Temporary or permanent passive seismic monitoring

"Short-term" applications mainly concern injection operations that are highly localised in time and space. The operator expects interpretation of the measurements immediately after the test, for hydraulic fracturing or injection control.

Monitoring of hydraulic fracture propagation (see 2.4.5) depends on the instruments implemented (number of sensors) and the knowledge available on wave propagation velocities in the different layers.

Injection control consists in checking that injection of a given fluid (production water, wastewater, industrial fluids, drilling mud, steam, CO_2, etc.) in a reservoir or an underground aquifer carried out in limited proportions (limited injection duration) proceeds correctly. Under these conditions, the operator does not want to fracture the geological units surrounding the reservoir layer or the aquifer, to avoid polluting the other layers.

Carried out near the injection zone, PSM can track possible fracturing of the rocks or reactivation of natural fractures or faults. If this type of phenomenon is observed, injection can be contained in the initial layer by simply adapting the injection parameters. On the contrary, if there is no microseismic activity and no brutal injection pressure change, there is a strong likelihood that the rock has not been fractured.

To date, "long-term" or permanent applications are operational in mining and storage of hydrocarbons in saline cavities. Their main role is to ensure the safety of people and property. These applications are also implemented for safety considerations in two other cases: oil production where permanent surface measures have already been imposed on some operators (especially NAM in the Netherlands) and storage of nuclear waste (NIREX).

In traditional oil and gas applications, long-term PSM proves interesting for gas storage in underground aquifers [Vidal *et al.*, 2000; Deflandre *et al.*, 2002], monitoring of oil production, monitoring of steam chamber propagation in thermal production [Snell *et al.*, 1999], reinjection of production water and more recently CO_2 sequestration.

Long-term applications can be expected to play an increasing role in meeting restrictions on protection of the environment [Snell *et al.*, 1999]. It is in fact likely that PSM will eventually be implemented in oil well abandonment procedures.

3.3.3.2 Surface PSM

A ground surface network of sensors can be used for detecting micro-seismic events. Surface PSM applications include fields for which measurements are taken on the surface and measurements that are easy to access (e.g. mine galleries). In practice, the same instruments are used.

In mine galleries, the main objective is to predict the risk of collapse during operation. In particular, the aim is to precisely locate the dislocation zones.

The oil industry may also be concerned by PSM surface applications. Further to the earthquakes of magnitude 5 on the Gutenberg-Richter scale, recorded around the Groningen field in the Netherlands, the government obliged the oil companies to implement surface monitoring to predict the seismic risk induced by massive extraction of hydrocarbons from the subsoil [Mobach and Gussinklo, 1994]. In this case, the major risk is failure of a dyke, which could have disastrous consequences. Studies have therefore been conducted to predict the maximum seismicity level that exploitation of the neighbouring fields could induce.

3.3.3.3 Downhole PSM

Given the low amplitude of microseismic signals and the uncertainties over wave propagation velocities in the surface layers, it became clear that buried sensor networks are recommended. More events can be recorded and located with greater accuracy. Furthermore, since the signals are less attenuated due to shorter trajectories, the frequency bandwidth is richer and, as a result, mechanical phenomena can be described with higher resolution.

Several types of instrument are used in wells: VSP, acoustic probes or permanent sensors (Figure 3.3.5).

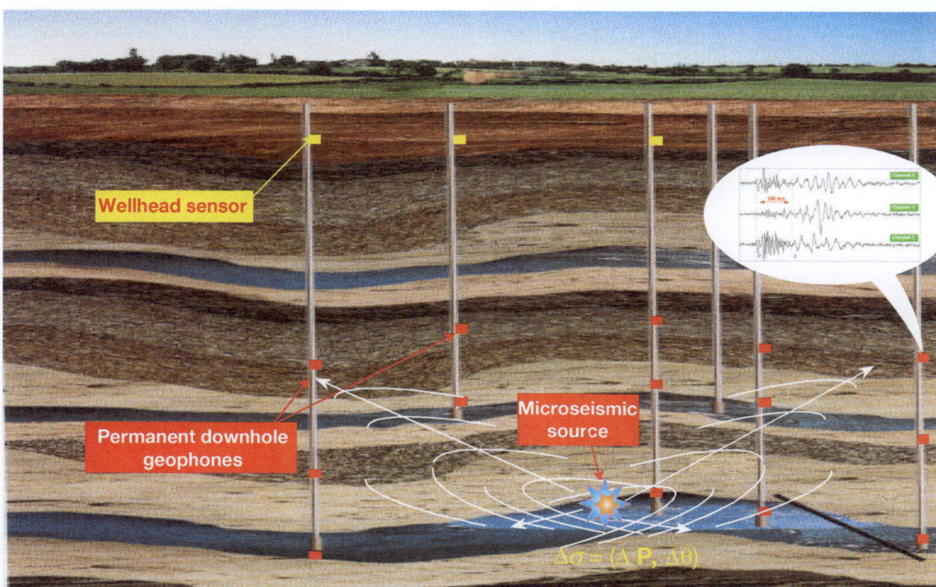

Figure 3.3.5

Principle of passive seismic monitoring with downhole instruments.

A. VSP (Vertical Seismic Profile) tools

The advantage with VSP tools is that an array of geophones is placed vertically in a wellbore. However, the instruments must be installed in a well dedicated to measurement (production stopped and production column removed) and the acquisition systems are not always suitable for permanent monitoring. On the Ekofisk and Valhall fields in the North Sea, records are taken by systematic triggering every 30 seconds (CGG-BSD seismic recording system) and a 2-second reset time is required. The microseismic events are detected when rereading the records. This method does not allow real-time monitoring.

B. Acoustic probes equipped with geophones

Several tools have been developed to monitor the recording of acoustic signals during hydraulic fracturing. These tools are mono- or multi-components. The tool developed by IFP Energies nouvelles (Figure 3.3.6) was designed for use in injection well and in observation well. The logging probe includes an acoustic module equipped with a triaxial accelerometer and a clock and an electronic module equipped with a pressure sensor and a temperature probe.

This system allows bottomhole fracturing temperature and pressure logging during and after injection and real-time logging of a large number of events due to the proximity of the emission zone. In addition, proximity of the emission zone reduces the uncertainty on the

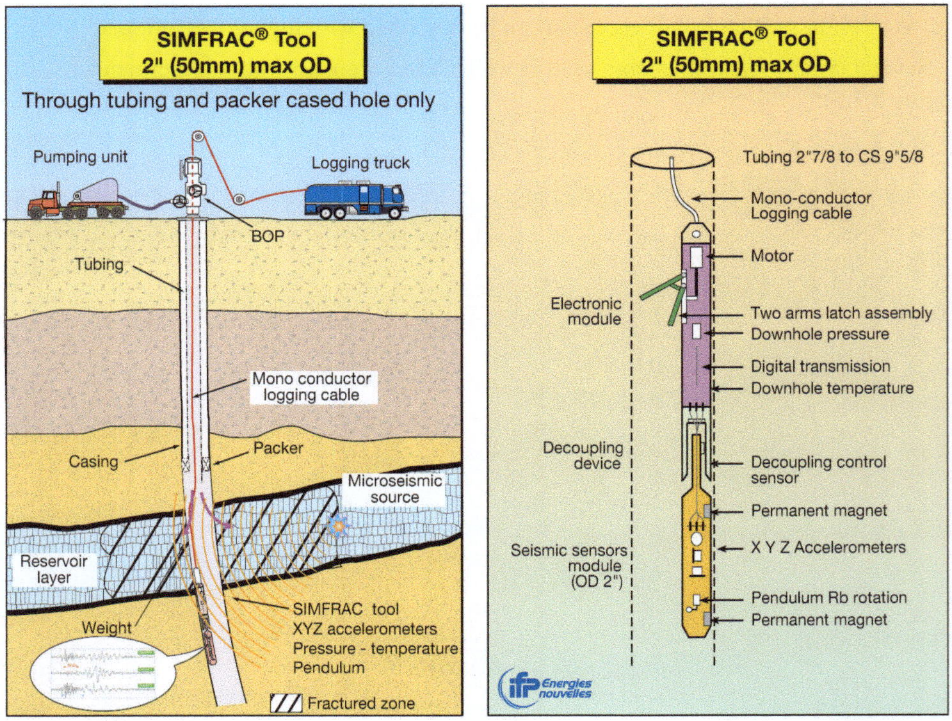

Figure 3.3.6

Diagram of the SIMFRAC logging probe during a fracturing experiment.

location of microseismic events and produces a more accurate map of the structure. Lastly, system installation costs are low since the tubing does not have to be pulled up to take the measurement.

In the frac-treated well, however, the microseismicity activity emitted during injection cannot be analysed due to the noise level in the well. The probe can only be used during the minifracturing phase preceding injection of the proppant.

Acoustic probes equipped with three geophones can be lowered down on a cable in a cased well not equipped with a production column. Coupling is provided by the opening of two motor-driven arms.

C. Well permanent seismic sensors

Passive seismic monitoring currently benefits from recent instrument developments with, in particular, well permanent seismic sensors and bottomhole numerical acquisition and transmission systems (Figure 3.3.7).

The big advantage with well permanent seismic sensors is that measurements can be taken without affecting production. This advantage will be increasingly important when operating deep sea wells due to their specificities (oil type, low pressure and temperature).

Comprehensive systems cover the requirements of conventional well seismics and passive seismic monitoring.

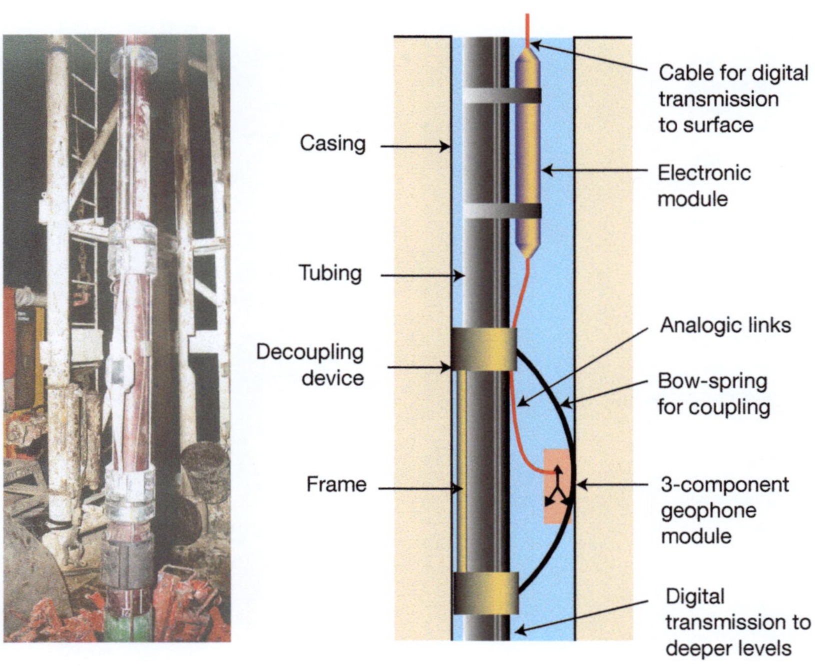

Figure 3.3.7

Well permanent seismic sensors [Laurent *et al.*, 1999].

References

Aadnoy, B.S. (1988) Inversion technique to determine the in situ stress field from fracturing data, Paper SPE 18023 presented at the 63rd Annual Technical Conference and Exhibition, Houston, Texas, October 2-5.

Aadnoy, B.S. and Chenevert, M.E. (1987) Stability of highly inclined boreholes, Paper SPE 16052 presented at the SPE/IADC Drilling Conference held in New Orleans, March 15-18.

Abaqus Analysis User's Manual, Abaqus Version 6.4, Simulia, Dassault Systèmes, 2005.

Abass, H.H., Hedayati, S. and Meadows, D.L. (1996) Nonplanar fracture propagation from a horizontal wellbore: experimental study, Paper SPE 24823, *SPE Production and Facilities*, pp. 133-137.

Addis, M. A. (1997) The stress-depletion response of reservoirs, Paper SPE 38720 presented at the SPE Annual Technical Conference and Exhibition held in San Antonio, Texas, October 5-8.

Addis, M.A., Choi, X. and Cumming, J. (1998) The influence of the reservoir stress-depletion response on the lifetime considerations of well completion design, Paper SPE 47289 presented at the SPE/ISRM Eurock'98 held in Trondheim, Norway, July 8-10.

Adenot, F. (1992) Durabilité du béton: Caractérisation des processus physiques et chimiques de dégradation du ciment, PhD. Thesis, Université d'Orléans.

Alfenore, J. (2001) How to control high productivity wells in deep water reservoirs, SPE Distinguished Lecturer Program.

Al-Harthy, S.S., Jing X.D., Marsden, J.R. and Dennis, J.W. (1999) Petrophysical properties of sandstones under true triaxial stresses I: Directional transport characteristics and pore volume change. Paper SPE 57287 presented at the SPE Asia Pacific Improved Oil Recovery Conference held in Kuala Lumpur, Malaysia, October 25-26.

Alonso, E. E., Gens, A. and Josa, A. (1990) A constitutive model for partially saturated soils, *Géotechnique*, Vol. 40, n°3, pp. 405-430.

Batzle, M. and Wang, Z. (1992) Seismic properties of pore fluids, *Geophysics*, Vol. 57, N° 11, pp. 1396-1408.

Baumgarte, C., Thiercelin, M. and Klaus, D. (1999) Case studies of expanding cement to prevent microannular formation, Paper SPE 56535 presented at the Annual Technical Conference and Exhibition, Houston, Texas, October 3-6.

Bear, J. and Buchlin, J.M. (1991) Modelling and applications of transport phenomena in porous media, Kluwer, Academic Publishers, Dordrecht.

Beaudoin, J.J. (2000) Calcium hydroxide in cement matrices: physico-mechanical and physico-chemical contributions, *In* Materials Science of Concrete: Calcium Hydroxide in Concrete, Special Volume, pp. 131-142.

Bemer, E., Boutéca, M., Vincké, O., Hoteit, N. and Ozanam, O. (2001) Poromechanics: from linear to nonlinear poroelasticity and poroviscoelasticity, *Oil & Gas Science and Technology – Rev. IFP*, Vol. 56, N° 6, pp. 531-544.

Bemer, E. and Lombard, J.-M. (2008) An experimental investigation of the evolution of rock poromechanical properties associated to chemical alteration processes, Proceedings of Third International Conference on Coupled T-H-M-C Processes in Geo-systems (GEOProc 08), Lille, France.

Berryman, J. G. (1995) Mixture theories for rock properties, in Rock physics and phase relations – A handbook of physical constants, edited by T. J. Ahrens, American Geophysical Union, Washington, pp. 205-228.

Bévillon, D. (2000) Couplage d'un modèle de gisement et d'un modèle mécanique: Application à l'étude de la compaction des réservoirs pétroliers et de la subsidence associée. PhD Thesis, *Université des Sciences et Technologies – Lille.*

Biot, M. A. (1941) General theory of three-dimensional consolidation, *Journal of Applied Physics*, Vol. 12, pp. 155-164.

Birch, F. (1961) The velocity of compressional waves in rocks to 10 kilobars, Part 2. *J. Geophys. Res.*, 66, pp. 2199-2224.

Boidy, E. (2002) Modélisation numérique du comportement différé des cavités souterraines, PhD Thesis, *Université Joseph Fourier – Grenoble 1.*

Bosma, M.G.R., Cornelissen, E.K. and Schwing, A. (2000) Improved experimental characterisation of cement/rubber zonal isolation materials, Paper SPE 64762 presented at the International Oil and Gas Conference and Exhibition in China held in Beijing, China, November 7-10.

Bourbié, T., Coussy, O. and Zinszner, B. (1986) Acoustique des milieux poreux, Editions Technip, Paris.

Boutéca, M., Lessi, J. and Sarda, J-P. (1983) Stress changes induced by fluid injection in a porous layer around a wellbore, Proceedings 24th US Symposium on Rock Mechanics, College Station, Texas.

Boutéca, M. and Sarda, J-P. (1987) Etat de l'art en fracturation hydraulique, *Oil & Gas Science and Technology – Rev. IFP*, Vol. 42, N° 1, pp. 39-75.

Boutéca, M. (1988) Hydraulic fracturing model based on a closed form : tests , analysis of fracture geometry and containment, *SPE Production Engineering*, Vol. 3, N° 4, pp. 445-454.

Boutéca, M. (1989) Three-dimensional hydraulic fracture propagation: derivation of closed forms from potential theory, Proceedings ISRM/SPE International Symposium on Rock at Great Depth, Vol. 2, pp. 971-979, Ed. V. Maury & D. Fourmaintraux, A.A. Balkema.

Boutéca, M. (1992) Elements of poro-elasticity for reservoir engineering, *Oil & Gas Science and Technology – Rev. IFP*, Vol. 47, N° 4, pp. 479-490.

Boutéca, M., Sarda, J.-P. and Schneider, F. (1996), Les subsidences provoquées par la production des fluides souterrains, *Oil & Gas Science and Technology – Rev. IFP*, Vol. 51, N° 3, pp. 349-364.

Boutéca, M.J., Sarda, J-P. and Vincké, O. (2000) Constitutive law for permeability evolution of sandstones during depletion, Paper SPE 58717 presented at the SPE International Symposium on Formation Damage Control held in Lafayette, Louisiana, February 23-24.

Bratli, R.K. and Risnes, R. (1981) Stability and failure of sand arches, Paper SPE 8427-PA, *SPE Journal*, Vol. 21, N° 2, pp. 236-248, April 1981.

Braun, R. and Jahns, E. (1998) Cracking and compaction behavior of sandstone, Paper SPE 47377 presented at the 1998 SPE/ISRM Conference on Rock Mechanics in Petroleum Engineering, Trondheim, Norway, July 8-10.

Brink, J.L., Patzek, T.W., Silin, D.B. and Fielding, E.J. (2002) Lost Hills field trial – Incorporating new technology for reservoir management, Paper SPE 77646 presented at the Annual Technical Conference and Exhibition held in San Antonio, Texas, 29 September-2 October.

Bull, W. B. (1964) Alluvial fans and near-surface subsidence in western Fresno County, California, USGS Professional Paper. 437-A, pp. 1-71.

Burger J., Gadelle, C.P. and Marrast, J.R. (1986) Development of a chemical process for sand control, SPE 15410 presented at the SPE Annual Technical Conference and Exhibition, New Orleans, Louisiana, October 5-8.

Byerlee, J.D. (1978) Friction of rocks, *Pure and Applied Geophysics*, Vol. 116, pp. 615-626.

Calvert, D.G. and Smith, D.K. (1994) Issues and techniques of plugging and abandonment of oil and gas wells, Paper SPE 28349 presented at the Annual Technical Conference and Exhibition, New Orleans, September 25-28.

Carde, C. (1996) Caractérisation et modélisation de l'altération des propriétés mécaniques due à la lixiviation des matériaux cimentaires, PhD INSA Toulouse.

Carter, B.J., Desroches, J., Ingraffea, A.R. and Wawrzynek, P.A. (2000) Simulating fully 3D hydraulic fracturing, *In* Modeling in Geomechanics, Ed Zaman, Booker and Gioda, Wiley Publishers.

Castagna, J.P., Batzle, M.L. and Eastwood, R.L. (1985) Relationships between compressional-wave and shear-wave velocities in clastic silicate rocks, *Geophysics*, Vol. 50, N° 4, pp. 571-581.

Cerasi, P., Papamichos, E. and Stenebraten, J. F. (2005) Quantitative sand-production prediction: Friction-dominated flow model, Paper 94791 presented at the SPE Latin American and Caribbean Petroleum Engineering Conference held in Rio de Janeiro, Brazil, June 20-23.

Chardac, O., Murray, D., Carnegie, A. and Marsden, J.R. (2005) A proposed data acquisition program for successful geomechanics projects, Paper SPE 93182 presented at the 14th SPE Middle East Oil and Gas Show and Conference, Bahrain, March 12-15.

Charlez, P.A. (1991) Rock mechanics, Volume 1, Theoretical fundamentals, Editions Technip, Paris.

Charlez, P.A. (1994) Exemple de modèle poroplastique: le modèle de Cam-Clay, Ecole de Mécanique des Milieux Poreux, Aussois, France, 30 Mai-3 Juin 1994.

Charlez, P.A., Lemonnier, P., Ruffet, C., Boutéca, M. and Tan, C. (1996) Thermally induced fracturing: analysis of a field case in North Sea, Paper SPE 36916 presented the SPE European Petroleum Conference held in Milan, Italy, October 22-24.

Charlez, P.A. (1997) Rock mechanics, Petroleum Applications. Editions Technip, Paris.

Chavez, J-C., Carruthers, J. and McMurdy, P. (2005) Water flooding efficiency in a scenario of multiple induced fractures, an applied geomechanical study, Paper SPE 97526 presented at the SPE International Improved Oil Recovery Conference in Asia Pacific held in Kuala Lumpur, Malaysia, December 5-6.

Chen, G., Chenevert, M., Sharma, M. and Yu, M. (2003) A study of wellbore stability in shales including poroelastic, chemical, and thermal effects, *Journal of Petroleum Science and Engineering*, Vol. 38, N° 3-4, June 2003, pp. 167-176.

Chen, X., Tan, C.P. and Detournay, C. (2003) A study on wellbore stability in fractured rock masses with impact of mud infiltration, *Journal of Petroleum Science and Engineering*, Vol. 38, N° 3, pp. 145-154.

Chi, J.M., Huang, R. and Yang, C.C. (2002) Effects of carbonation on the mechanical properties and durability of concrete using accelerated testing method, *Journal of Marine Science and Technology*, Vol. 10, No. 1, pp. 14-20.

Chin, L.Y. and Thomas, L.K. (1999) Fully coupled analysis of improved oil recovery by reservoir compaction, Paper SPE 56753 presented at the SPE Annual Technical Conference and Exhibition, Houston, Texas, October 3-6.

Christensen, O., Janbu, N. and Jones, M.E. (1988) Subsidence due to oil-gas production, Proceedings of the International Conference on Behaviour of Offshore Structures (BOSS' 88), Trondheim, Norway, V1, pp. 143-157.

Cipolla, C.L. and Wrigth, C.A. (2002) Diagnostic techniques to understand hydraulic fracturing: what? Why? and How?, Paper SPE 75359, *SPE Production & Facilities*, Vol. 17, N° 1, pp. 23-35.

Cook, J. M., Sheppard, M. C. and Houwen, O. H. (1991) Effects of strain rate and confining pressure on the deformation and failure of shale. Paper SPE 19944 presented at the 1990 IADC/SPE Drilling Conference held in Houston, Texas, February 27-March 2.

Cooper, G. A. and Hatherly, P. (2003) Prediction of rock mechanical properties from wireline data and their use in drilling simulation, Paper SPE 83509 presented at the SPE Western Regional/AAPG Pacific Section Joint Meeting held in Long Beach, California, May19-24.

Cossé, R. (1993) Basic of reservoir engineering, Editions Technip, Paris.

Coussy, O. (1991) Mécanique des milieux poreux, Editions Technip, Paris.

Coussy, O. (1995) Mechanics of Porous Continua, John Wiley & Sons Ltd., Chichester.

Coussy, O. (2004) Poromechanics, John Wiley & Sons Ltd., Chichester.

Crawshaw, J. P. and Frigaard, I. (1999) Cement plugs: stability and failure by buoyancy-driven mechanism, Paper SPE 56959 presented at the Offshore Europe Conference held in Aberdeen, Scotland, September 7-9.

Cunningham, R.A. (1978) An empirical approach for relating drilling parameters, Paper SPE 6715, *Journal of Petroleum Technology*, Vol. 30, N° 7, pp. 987-991, July 1978

Damjanac, B. and Detournay, E. (1995) Numerical modeling of normal wedge indentation in rocks, *in Rock Mechanics*, Proceedings of the 35th U.S. Symposium, University of Nevada, Reno, pp. 349-354. J.J.K. Daemen and R. A. Schultz, Balkema Editions, Rotterdam.

Dangla, P., Malinski, L. and Coussy, O. (1997) Plasticity and imbibition-drainage curves for unsaturated soils: A unified approach, Proc. of the 6th International Symposium on Numerical Models in Geomechanics (NUMOGVI), Montréal, Quebec, July 2-4, Edited by S. Pietruszczak and G. Pande. A.A. Balkema, Rotterdam, the Netherlands. pp. 141-147.

De Borst, R. and Nauta, P. (1985) Non-orthogonal cracks in a smeared finite element model. *Engineering Computations*, Vol. 2, N° 1, pp. 35-46.

De Sa, A. N., Tavares, A. F. C. and do Carmo Marques (1989) Gravel Pack in offshore wells, Proc. of the 21th Annual Offshore Technology Conference, Houston, paper 6041.

De Waal, J.A. and Smits, R.M.M. (1988) Prediction of reservoir compaction and surface, Paper SPE 14214, SPE Formation Evaluation, Vol. 3, N° 2, pp. 340-346.

Deflandre, J-P. and Huguet, F. (2002) Microseismic monitoring on gas storage reservoirs: a ten-year experience, Proc. 17th World Petroleum Congress, Rio de Janeiro, Brazil.

Deliac, E (1986) Optimisation des machines d'abattage à pics, PhD Thesis, *Université Pierre et Marie Curie.*

Desroches, J. and Kurkjian, A.L. (1998) Applications of wireline stress measurements, SPE 48960 presented at the SPE Annual Technical Conference and Exhibition, New Orleans, September 27-30.

Desroches, J., Marsden, J.R. and Colley, N., M. (1998) Wireline open-hole stress tests in a tight gas sandstone, Proceedings of the International Gas Research Conference, Cannes, France, November 6-9.

Detienne, J-L., Danquigny, J., Lacourie, Y. and Espy, M. (2002) Produced water re-injection on a low permeability carbonaceous reservoir, Paper SPE 78482 presented at the 10th Abu Dhabi International Petroleum Exhibition and Conference, October 13-16.

Detournay, E. and Atkinson, C. (1991) Influence of pore pressure on the drilling response of PDC bits, Rock Mechanics as a Multidisciplinary Science, Ed. by J.C. Roegiers, Balkema, Rotterdam, pp. 539-547.

Detournay, E. and Defourny, P. (1992) A phenomenological model for the drilling action of drag bits, *International Journal of Rock Mechanics and Mining Sciences & Geomechanics*, Vol. 29, N° 1, pp. 13-23.

Detournay, E. (1996) Models for rock cutting, Proceedings of the Tool-Rock Interaction Workshop held during the 2nd North American Rock Mechanics Symposium (NARMS-2), Montreal, June 22, Editions Balkema.

Dubucq, D., Chalier, G. and Xavier, J-P. (2008) Steam chamber monitoring using surface deformation from radar (PS-InSAR) and tiltmeter measurements, Proceedings of the Canadian CSPG CSEG CWLS Convention.

Dusseault, M. B. and Santarelli, F. J. (1989) A conceptual model for massive solids production in poorly consolidated sandstones, Proceedings ISRM/SPE International Symposium on Rock at Great Depth, Vol. 2, Balkema, pp. 789-798.

Dusseault, M, Geilickman, M. and Spanos, T. (1998) Mechanisms of sand production in heavy oils, Proceedings 7th UNITAR Conference on Heavy Oil and Tar Sands, Beijing, N° 1998.123.

Economides, M.J., Watters, L.T. and Dunn-Norman, S. (1998) Petroleum well construction, John Wiley & Sons.

Egermann, P., Bemer, E. and Zinszner, B. (2006) An experimental investigation of the rock properties evolution associated to different levels of CO_2 injection like alteration processes, Paper SCA 2006-34, Proceedings of the International Symposium of the Society of Core Analysts, Trondheim, Norway, September 12-16.

Eilers, L.H. and Root, R.L. (1974) Long-term effects of high temperature on strength retrogression of cements, SPE, Paper 5028 presented at the 49th Annual Fall Meeting of the SPE of AIME, Houston, Texas, USA, October 6-9.

Elkins, L., Morton, D. and Blackwell, W. (1972) Experimental fireflood in a very viscous oil-consolidated sand reservoir, Paper SPE 4086 presented at the Fall Meeting of the Society of Petroleum Engineers of AIME, 8-11 October 1972, San Antonio, Texas.

Ellis, R.C. (1998) An overview of frac packs: a technical revolution (evolution) process, Journal of Petroleum Technology, Vol. 50, N° 1, pp 66-68.

Enever, J.R., Yassir, N., Willoughby, D.R. and Addis, M.A. (1996) Recent experience with extended leak-off tests for in situ stress measurements in Australia, *Australian Petroleum Production and Exploration Association Journal*, pp. 528-535.

Engstrøm, F. (1991) Rock mechanical properties of danish North Sea Chalk, Proceedings of the Fourth North Sea Chalk Symposium. Deauville, France, September 21-23.

Eronini, I. E., Somerton, W. H. and Auslander, D. M. (1982) A Dynamic model for rotary rock drilling, Transactions of ASME, *Journal of Energy Resources Technology,* Vol. 104, N° 2, pp. 108-120.

Evans, G.W. and Carter, L.G. (1963) Bonding studies of cementing compositions to pipe and formations, drilling and production practice 1962, American Petroleum Institute, New York.

Falconer, I.G., Burgess, T.M. and Sheppard, M.C. (1988) Separating bit and lithology effects from drilling mechanics data, Paper SPE 17191 presented at the SPE/IADC Drilling Conference, Dallas, Texas, February 28 -March 2.

Fam, W.A. and Dusseault, M.B. (1998) Borehole stability in shales: a physico-chemical perspective, Paper SPE 47301 presented at the SPE/ISRM Conference on Rock Mechanics in Petroleum Engineering, Eurock'98, Trondheim, Norway, Balkema, Rotterdam, pp. 285-290.

Fear, M.J., Knowlton, R.H., Meany, N. and Warren, T.M. (1995) Experts discuss drill-bit design, field performance, *Journal of Petroleum Technology*, 100-104, pp. 142-143.

Finol, A.S. and Sancevic, Z. A., (1995) Subsidence in Venezuela, in subsidence due to fluid withdrawal, developments in petroleum science, edited by G. V. Chilingarian, E. C. Donaldson and T.F. Yen, Elsevier Science, Amsterdam.

Fjær, E., Holt, R. M., Horsrud, P., Raaen, A. M. and Risnes, R. (1992) Petroleum related rock mechanics, Elsevier, Amsterdam.

Franco, J.L.A., De La Torre, H.G., Ortiz, M.A.M., Wielemaker, E., Plona, T.J., Saldungaray, P. and Mikhaltseva, I. (2005) Using shear-wave anisotropy to optimize reservoir drainage and improve production in low-permeability formations in the North of Mexico, Paper SPE 96808 presented at the SPE Annual Technical Conference and Exhibition held in Dallas, Texas, October 9-12.

Fung L. S. K., Buchanan, L., and Wan, R. G. (1992) Coupled geomechanical-thermal simulation for deforming heavy-oil reservoirs. Proceedings of the CIM Annual Technical Conference, Calgary, June 7-10.

Galloway, D.L., Hudnut, K.W., Ingebritsen, S.E., Phillips, S.P., Peltzer, G., Rogez, F. and Rosen, P.A. (1998) InSAR detection of aquifer-system compaction and land subsidence, Antelope Valley, Mojave Desert, California, Water Resources Research., Vol. 34, 2573-2585.

Galloway, D.L., Jones, D.R. and Ingebritsen, S.E. (1999) Land subsidence in the United States: U.S. Geological Survey Circular 1182.

Garnier, A.J. and van Lingen, N.H. (1958) Phenomena affecting drilling rates at depth, Paper SPE 1097 presented at 33rd SPE Annual Meeting, Houston, in Petroleum Transactions, AIME, Volume 216, 1959, pages 232-239.

Gassmann, F. (1951) Über die Elastizität Poröser Medien, Veirteljahrsschrift der Naturforschenden Gesellschaft, in Zurich, 96, 1-23.

Gatto, P. and Carbognin, L. (1981) The Lagoon of Venice: natural environmental trend and man-induced modification, *Hydrological Sciences Bulletin*, 26/4/12, pp. 379-391.

Gavito, D. G. (1996) A new rock strength model and its practical applications, Paper SPE 35322 presented at the International Petroleum Conference and Exhibition of Mexico City, March 5-7.

Geertsma, J. (1973) A basic theory of subsidence due to reservoir compaction: the homogeneous case, *Verhandelingen van het Koninklijk Nederlands Geologisch Mijnbouwkundig Genootschap,* Vol. 28, pp. 43-62.

Geertsma, J., and Van Opstal, G. (1973) A numerical technique for predicting subsidence above compacting reservoirs, based on the nucleus of strain concept, *Verhandlingen van het Koniklijk Nederlands Geologisch Mijnbouwkundig Genootschap*, Vol. 28, pp. 63-78.

Geertsma, J. (1989) On the alert for subsidence, *Agip Review*, Vol. 6, pp. 39-43.

Geilickman, M. and Dusseault, M. (1997) Dynamics of wormholes and enhancement of fluid production, Paper 97-09 presented at the 48th Annual Technical Meeting of the Petroleum Society in Calgary, Canada.

Geilickman, M. Dusseault, M. and Dullien, F.A. (1994) Fluid saturated solid flow with propagation of a yielding front, Proceedings of the SPE/ISRM Conference on Rock Mechanics in Petroleum Engineering, Eurock'94, Delft, Balkema, Rotterdam.

Geoffroy, H. (1996) Etude de l'interaction roche-outil de forage: influence de l'usure sur les paramètres de coupe, PhD thesis, Ecole Polytechnique, Palaiseau, France.

Geoffroy, H., Nguyen Minh, D. and Putot, C. (1997) Study on interaction between rocks and worn PDC's cutter, Proceedings of the 36th Symposium US on Rock Mechanics ISRM, New York.

Geoffroy, H., Nguyen Minh, D., Bergues, J. and Putot, C. (1998a) Frictional contact on cutters wear flat and evaluation of drilling parameters of a PDC bit, Paper SPE 47323 presented at the SPE/ISRM Conference on Rock Mechanics in Petroleum Engineering, Trondheim, Norway, July 8-10.

Geoffroy, H., Nguyen Minh, D., Maitournam, H., Bergues, J. and Putot, C. (1998b) Evaluation of drilling parameters of a PDC bit, in Advances in Rock Mechanics, World Scientific Publishing Co Ed., pp. 301-314.

Gérard, B. (1996) Contribution des couplages mécanique-chimie-transfert dans la tenue à long terme des ouvrages de stockage des déchets radioactifs, PhD Thesis, *Ecole Normale Supérieure*, Cachan, France.

Gidley, J.L., Holditch, S.A., Nierode, D.E. and Veatch, R.W. (1989) Recent advances in hydraulic fracturing, Society of Petroleum Engineers, Richardson, TX, USA., Monograph SPE Vol. 12.

Glowka, D. A. (1989) Use of single-cutter data in the analysis of PDC bit designs: Part 1 – Development of a PDC cutting force model. Part 2 – Development and use of the PDCWEAR computer code, *Journal of Petroleum Technology*, August. Vol. 41, N° 8.

Glowka, D.A. (1987) Development of a method for predicting the performance and wear of PDC drill bits, Sandia Report 86-1745, Sandia National Laboratories.

Green, S. J., Griffin, R. M. and Pratt, H. R. (1973) Stress-strain and failure properties of a porous shale. Paper SPE 4242 presented at the SPE Drilling and Rock Mechanics Conference, Austin, Texas, January 22-23.

Guéguen, Y. et Palciauskas, V. (1992) Introduction à la physique des roches, Editions Hermann, Paris.

Guéguen, Y. and Boutéca, M. (1999), Mechanical properties of rocks: Pore pressure and scale effects. *Oil & Gas Science and Technology – Rev. IFP*, Vol. 54, N° 6, pp. 703-714.

Guillemot, J. (1986) Elements de géologie, Editions Technip, Paris.

Gutierrez, M. and Lewis, M. (1998) The role of geomechanics in reservoir simulation, Proceedings of the SPE/ISRM Conference on Rock Mechanics in Petroleum Engineering, Trondheim, Norway, Eurock'98, Trondheim, Norway, July 8-10, pp. 439-448.

Hanson, J.M. (1991) Pore pressure ahead of the bit, Paper SPE 21916 presented at SPE/IADC Drilling Conference, Amsterdam, Netherlands, March 11-14.

Hatchell, P.J., van den Beukel, A., Molenaar, M.M., Maron, K.P., Kenter, C., Stammeijer, J.G.F., van der Velde, J.J. and Syers, C.M. (2003) Whole earth 4D: reservoir monitoring geomechanics, Expanded Abstracts of the 73rd Annual International Meeting of the Society of Exploration Geophysicists, Dallas, pp. 1330-1333.

Hawkes, C.D. and McLellan, P.J. (1997) A new model for predicting time-dependent failure of shales: theory and application, Paper presented at the 48th Technical Meeting of the Petroleum Society of CIM, Calgary, June, also published in the Journal of Canadian Petroleum Technology, Vol. 30, N° 13, December 1999.

Hawkes, C.D., McLellan, P.J. and Bachu, S. (2005) Geomechanical factors affecting geological storage of CO_2 in depleted oil and gas reservoirs, *Journal of Canadian Petroleum Technology* vol. 44, N° 10, pp. 52-61.

Hebert, W.E. and Young, F.S. (1972) Estimation of formation pressure with regression models of drilling data, *Journal of Petroleum Technology*, Volume 24, Number 1, pp. 9-15.

Heffer, K.J., Last, N.C., Koutsabeloulis, N.C., Chan, H.C.M., Gutierrez, M. and Makurat, A. (1994) The influence of natural fractures, faults and earth stresses on reservoir performance – Geomechanical analysis by numerical modelling. North Sea oil and gas reservoirs – *Norwegian Geotechnical Institute* (Revue), Vol. 192, pp. 201-211.

Hermansen, H., Landa, G.H., Sylte, J.E. and Thomas L.K. (2000) Experiences after 10 years of waterflooding the Ekofisk Field, Norway. *Journal of Petroleum Science and Engineering,* Vol. 26, N° 1-4, pp. 11-18.

Hettema, M.H.H., Schutjens P.M.T.M., Verboom B.J.M. and Gussinklo H.J. (1998) Production-induced compaction of sandstone reservoirs: The strong influence of field stress, Paper SPE 50630 presented at the European Petroleum Conference, 20-22 October 1998, The Hague, Netherlands.

Hettema, M., Papamichos, E. and Schutjens, P. (2002) Subsidence delay: field observations and analysis, *Oil & Gas Science and Technology – Rev. IFP*, Vol. 57, N° 5, pp. 443-458.

Homand, S. (2000) Comportement mécanique d'une craie très poreuse avec prise en compte de l'effet de l'eau: de l'expérience à la simulation, PhD Thesis, University of Lille.

Horsrud, P. (2001) Estimating mechanical properties of shale from empirical correlations, Paper SPE 56017, *Journal SPE Drilling & Completion,* Vol. 16, N° 2, pp. 68-73.

Horsrud, P., Holt, R., Sonstebo, E.F., Svano, G. and Bostrom, B. (1994) Time dependent borehole stability: laboratory studies and numerical simulation of different mechanisms in shale, Paper SPE 28060 presented at the Conference on Rock Mechanics in Petroleum Engineering, Eurock 94, 29-31 August, Delft, Netherlands.

Hu, T., Fournier, F., Royer, J.J. and Joseph, C. (2006) Impact of uncertainties in rock mechanical properties on subsidence evaluation, Proc. of the XIth International Congress of the International Association for Mathematical Geology, University of Liège – Belgium.

Huang, W.S., Marcum, B.E., Chase, M.R. and Yu, C.L. (1998) Cold production of heavy oil from horizontal wells in the Frog Lake field, Paper SPE 52636, *Journal SPE Reservoir Evaluation & Engineering*, Vol. 1, N° 6, pp. 551-555.

Jaeger, J. C. and Cook, N. G. W. (1979) Fundamental of rock mechanics – Third Edition, Chapman and Hall, London.

Jardine, S., Malone, D. and Sheppard, M. (1994) Putting a damper on drilling's bad vibrations, *Oilfield Review*, January 1994.

Johnson, K.L (1985) Contact mechanics, Cambridge University Press.

Johnson, J.P., Rhett, D.W. and Siemers, W.T. (1989) Rock mechanics of the Ekofisk reservoir in the evaluation of subsidence, Paper SPE 17854, *Journal of Petroleum Technology*, Vol. 41, N° 7, pp. 717-722.

Kadri, I. B. (1991) Abnormal formation pressures in post-eocene formation, Potwar Basin, Pakistan, Paper SPE 21920 presented at the SPE/IADC Drilling Conference, Amsterdam, Netherlands, 11-14 March.

Keaney, G.M.J., Meredith, P.G. and Murell, S.A.F. (1998) Laboratory study of permeability evolution in a 'tight' sandstone under non-hydrostatic stress conditions. Paper SPE 47265 presented at the SPE/ISRM Conference on Rock Mechanics in Petroleum Engineering, Trondheim, Norway, 8-10 July.

Kelm, C. H. and Faul, R. R. (1999) Well abandonment – A "best practices" approach can reduce environmental risk, Paper SPE 54344 presented at the SPE Asia Pacific Oil and Gas Conference and Exhibition, Jakarta, Malaysia, April 20-22.

Kendall, K. (1978) Complexities of compression failure, Proceedings of the Royal Society of London. Series A, 361, pp. 245-263.

Kenter, C.J., Blanton, T.L., Schreppers, G.M.A., Baaijens, M.N. and Ramos, G.G. (1998) Compaction study for Shearwater Field, Paper SPE 47280 presented at the SPE/ISRM Conference on Rock Mechanics in Petroleum Engineering, Trondheim, Norway, July 8-10.

Klimentos, T., Badri, M., Hassan, S. and Zeid, A. (1999) Determination of earth stresses, azimuthal rock-mechanical properties and fracture permeability using dipole shear sonic logging in the Gulf of Suez, Egypt, Proceedings of the 40th Annual SPWLA Logging Symposium, Oslo, Norway, May 30-June 3.

Koutsabeloulis, N.C., Heffer, K.J. and Wong, S. (1994) Numerical geomechanics in reservoir engineering, Computer methods and advances in geomechanics, Siriwardane and Zaman Editions, Balkema, Rotterdam.

Koutsabeloulis, N.C. and Hope S. A. (1998) Coupled stress/fluid/thermal multi-phase reservoir simulation studies incorporating rock mechanics, Paper SPE 47393 presented at the SPE/ISRM Conference on Rock Mechanics in Petroleum Engineering, Trondheim, Norway, July 8-10, pp. 449-454.

Kovary, K., Tisa, A., Einstein, H. H. and Franklin, J. A. (1983) Suggested methods for determining the strength of rock materials in triaxial compression: revised version. *Int. J. Rock Mech. Min. Sci. & Geomech. Abstr.*, Vol. 20, No.6, pp. 283-290.

Kuru, E. and Wojtanowicz, A.K. (1988) A method for detecting in-situ PDC bit dull and lithology change, Paper SPE 17192 presented at the SPE/IADC Drilling Conference, Dallas, Texas, 28 February-2 March 1988.

Lacy, L.L. (1997) Dynamic rock mechanics testing for optimized fracture designs, Paper SPE 38716 presented at the SPE Annual Technical Conference and Exhibition, San Antonio, Texas, October 5-8.

Lambe, T.W. and Whitman, R.W. (1969) Soil Mechanics, John Wiley &Sons, Inc, New York.

Landro, M. (1999) Discrimination between pressure and fluid saturation changes from time-lapse seismic data, 69th Annual International Meeting, SEG, Expanded Abstracts, pp. 1651-1654.

Last, N., Plumb, R., Harkness, R., Charlez, P., Alsen, J. and McLean, M. (1995) An integrated approach to evaluating and managing wellbore instability in the Cuisiana field, Colombia, South America, Paper SPE 30464 presented at SPE Annual Technical Conference and Exhibition held in Dallas, Texas, October 22-25.

Laurent, J., Deflandre, J.P, Dubois, J.C. and Huguet, F. (1999) Permanent Downhole Geophones: an efficient approach for reservoir monitoring, Proceedings of the European Applied Research Conference on Natural Gas 1999 ("Eurogas 99") Bochum, Germany.

Lavergne, M. (1986) Méthodes sismiques, Editions Technip, Paris.

Lawn, B.R. and Wilshaw, T.R. (1975) Fracture of brittle solids, Cambridge University Press.

Lesso, J.R. and Burgess, T.M. (1986) Pore pressure and porosity from MWD measurements, Paper SPE 14801 presented at the SPE/IADC Drilling Conference, Dallas, Texas, February 9-12.

Le Tirant, P. and Gay, L. (1972) Manuel de fracturation hydraulique, Editions Technip, Paris.

Lewis, R. W. and Schrefler, B.A. (1998) The Finite element method in the static and dynamic deformation and consolidation of porous media, Second Edition, Wiley & Sons, New York.

Lin, W. (1981) Mechanical behavior of Mesaverde shale and sandstone at high pressure, Paper SPE 9835 presented at the SPE/DOE Low Permeability Gas Reservoirs Symposium, Denver, Colorado, May 27-29.

Lofgren, B. E. (1975) Land subsidence due to ground-water withdrawal, Arvin-Maricopa area, California, USGS Professional Paper 437-D, Washington, D.C.

Longuemare, P., Détienne, J-L., Lemonnier, P., Boutéca, M. and Onaisi, A. (2001) Numerical modeling of fracture propagation induced by water injection/re-injection, Paper SPE 68974 presented at the SPE European Formation Damage Conference held in The Hague, The Netherlands, May 21-22.

Longuemare, P., Mainguy, M., Lemonnier, P., Onaisi, A., Gérard, Ch. and Koutsabeloulis, N. (2002) Geomechanics in reservoir simulation: overview of coupling methods and field case study, *Oil & Gas Science and Technology – Rev. IFP*, Vol. 57, No. 5, pp. 471-483.

Lumley, D. E. (1995) Seismic time-lapse monitoring of subsurface fluid flow, PhD. Thesis, Stanford University.

Mainguy, M. and Longuemare, P. (2002) Coupling fluid flow and rock mechanics: formulations of the partial coupling between reservoir and geomechanical simulators, *Oil & Gas Science and Technology – Rev. IFP*, Vol. 57, N° 4, pp. 355-367.

Mainguy, M., Longuemare, P., Audibert, A. and Lécolier, E. (2007) Analysing the risk of well plug failure after abandonment, *Oil & Gas Science and Technology – Rev. IFP*, Vol. 62, N° 3, pp. 311-324.

Marsden, S.S., and Davis, S.N. (1967) Geological subsidence, *Scientific American*, 216(6), pp. 93-108

Matà, C. (2001) Etude expérimentale et modélisation mécanique des effets du balayage à l'eau dans une craie saturée d'huile – Application à l'industrie pétrolière. *PhD Thesis*, Ecole Nationale des Ponts et Chaussées, Paris.

Maury, V. and Sauzay, J.M. (1987) Borehole instability: case histories, rock mechanics approach and results, Paper SPE 16051 presented at the SPE/IADC Drilling Conference, New Orleans, Louisiana, March 15-18.

Maury, V. (1994) Rock mechanics over the last ten years in oil and gas production: safety and economy, 14th World Petroleum Congress, Stavanger, Norway, Paper N° 26135.

Maury, V., Piau, J.-M. and Halle, G. (1996) Subsidence induced by water injection in water sensitive reservoir rocks: the example of Ekofisk, Paper SPE 36890 presented at the SPE European Petroleum Conference held in Milan, Italy, October 22-24.

McLean, M.R. and Addis, M.A. (1990) Wellbore stability: the effect of strength criteria on mud weight recommandations, Paper SPE 20405 presented at the 65th SPE Annual Technical Conference and Exhibition, New Orleans, September 23-26.

McLellan, P. (1994) Assessing the risk of wellbore instability in horizontal and inclined wells, Paper HWC94-14 presented at the Canadian SPE/CIM/CANMET International Conference on Recent Advances in Horizontal Well Applications, Calgary, Canada, March 20-23.

McLellan, P.J. and Cormier, K. (1996) Borehole instability in fissile, dipping shales, northeastern British Columbia, Paper SPE 35634 presented at the SPE Gas Technology Symposium, Calgary, Alberta, Canada, 28 April-1 May.

Mendelsohn, D.A. (1984) A review of hydraulic fracture modeling, part 1: general concepts, 2D models, motivation for 3D modeling, *Journal of Energy Resources Technology*, Vol. 106, N° 3, pp. 369-376.

Mendelsohn, D.A. (1984) A review of hydraulic fracture modeling, part II: 3D modeling and vertical growth in layered rock, *Journal of Energy Resources Technology*, Dec 1984, Vol. 106, N° 4, pp. 543-553.

Merle, H.A., Kentie, C.J.P., van Opstal, G.H.C. and Schneider, G.M.G. (1976) The Bachaquero study – A composite analysis of the behavior of a compaction drive/solution gas drive reservoir, *Journal of Petroleum Technology*, September, Vol. 28, N° 9, pp. 1107-1115.

Mindlin, R. D. (1949) Compliance of elastic bodies in contact, *Journal of Applied Mechanics*, Vol. 16, pp. 259-262.

Minkoff, S. E., Stone, C. M., Bryant, S. and Peszynska M. (2004) Coupled geomechanics and flow simulation for time-lapse seismic modelling, *Geophysics*, Vol. 69, N° 1, pp. 200-211.

Mobach, E. and Gussinklo, H. J. (1994) *In* situ reservoir compaction monitoring in the Groningen field, Proceedings of EUROCK'94, Rock Mechanics in Petroleum Engineering, A. A. Balkema Publ., Rotterdam, The Netherlands, pp. 535-547.

Moricca, G. Ripa, G., Sanfilippo, F. and Santarelli, F.J. (1994) Basin scale rock mechanics: field observations of sand production, Paper SPE 28066 presented at the SPE/ISRM Rock Mechanics in Petroleum Engineering Conference Delft, The Netherlands, August 29-31.

Morita, N., Whitfill, D.L., Nygaard, O. and Bale, A. (1989) A quick method to determine subsidence, compaction and in-situ stresses induced by reservoir depletion, Journal of Petroleum Technology, January 1989, SPE 17150, Vol. 41, N° 1, pp. 71-83.

Mouchet, J. P. and Mitchell, A. (1989). Abnormal pressure while drilling. Manuals techniques 2, Boussens, France, Elf Aquitaine Editions.

Mouton, D.E., Wilton, B.S. and Ramos, G.G. (1998) Optimizing drilling strategies in a tectonic belt, Pagerungan field, North of Bali, Paper SPE 39357 presented at the IADC/SPE Drilling Conference, Dallas, Texas, March 3-6.

Nguyen, J.P. (1993) Le forage, Editions Technip, Paris.

Nguyen Minh, D. (1974) Contribution à l'étude de la taille des roches, PhD thesis, University of Paris VI.

Noik, C. and Rivereau, A. (1999) Oilwell cement durability, Paper SPE 56538 presented at the SPE Annual Technical Conference and Exhibition, Houston, Texas, October 3-6.

Nolte, K.G. (1979) Determination of fracture parameters from fracturing pressure decline, Paper SPE 8341 presented at the SPE Annual Technical Conference and Exhibition, Las Vegas, Nevada, September 23-26.

Nolte, K.G. and Smith, M.B. (1981) Interpretation of fracturing pressures, *Journal of Petroleum Technology*, Vol. 33, N° 9, Sept 1981, pp. 1767-1775.

Nouri, A., Vaziri, H., Belhaj H. and Islam, R. (2004) Sand production control: a new set of criteria for modeling based on large scale transient experiments and numerical investigation, *SPE Jounal*, Vol. 11, N° 2, pp. 227-237.

Nouri, A., Vaziri, H., Kuru, E. and Islam, R. (2006) A comparison of two sanding criteria in physical and numerical modeling of sand production, *Journal of Petroleum Science and Engineering*, Vol. 50, N° 1, pp. 55-70.

Nur, A. and Murphy, W. (1981) Wave velocities and attenuation in porous media with fluids, Proceedings of the 4th International Conference on Continuum Models of Discrete Systems, Stockholm, pp. 311-327.

Onaisi, A., Durand, C. and Audibert, A. (1994) Role of hydration state of shales in borehole stability studies, Paper SPE 28062 presented at the SPE/ISRM Conference on Rock Mechanics in Petroleum Engineering, Eurock'94, Delft, Netherlands.

Osorio, J. G., Chen, H.Y., Teufel, L. and Schaffer, S. (1998) A two-domain, fully coupled fluid-flow/ Geomechanical simulation model for reservoirs with stress-sensitive mechanical and fluid-flow properties. Paper SPE 28062 presented at the SPE/ISRM Conference on Rock Mechanics in Petroleum Engineering, Eurock'98, Trondheim, Norway, July 8-10.

Papamichos, E., Stenebraten, J., Cerasi, P., Lavrov, A., Vardoulakis, I., Fuh, G.H., Brignoli, M., Goncalves de Castro, C.J. and Havmoller, O. (2008) Rock type and hole failure pattern effects on sand production, Proceedings of the 42nd US Rock Mechanics Symposium, San Francisco, June 29-July 2, Paper ARMA 08-217.

Patron, E. (1999) Porомécanique des joints en graphite expansé matricé, PhD Thesis, Ecole Nationale des Ponts et Chaussées, Paris.

Parcevaux, P. A. and Sault, P. H. (1984) Cement shrinkage and elasticity: a new approach for a good zonal isolation, Paper SPE 13176 presented at the 59th SPE Annual Technical Conference and Exhibition, Houston, September 16-19.

Pavone, D. R. and Desplans, J. P. (1994) Application of sampling rate downhole measurements for analysis and cure of stick-slip in drilling, Paper SPE presented at the SPE Annual Technical Conference and Exhibition, New Orleans, Louisiana, September 25-28.

Peltier, B. and Atkinson, C. (1986) Dynamic pore pressure ahead of the bit, Paper SPE 14787, *SPE Drilling Engineering*, Vol. 2, N° 4, pp. 351-358.

Perkins, T. K. and Weingarten (1988) Stability and failure of spherical cavities in unconsolitated sand and weakly consolidated rock, Paper SPE 18244 presented at SPE Annual Technical Conference and Exhibition, Houston, Texas, October 2-5.

Perkins, T.K. and Gonzalez, J.A. (1985) The effect of thermoelastic stressses on injection well fracturing, Paper SPE 11332, *SPE Journal*, Vol. 25, N° 1, pp. 78-88.

Perrodon, A. (1989) Dynamique et méthode d'étude des bassins sédimentaires, Editions Technip, Paris.

Pessier, R.C (1994) Different shales dictate fundamentally different strategies in hydraulics, bit selection and operating practices, Paper SPE 28322 presented at the 69th Annual Technical Conference and Exhibition, New Orleans, Louisiana, September 25-28.

Pessier, R. C. and Fear, M. J. (1992) Quantifying common drilling problems with mechanical specific energy and a bit-specific coefficient of sliding friction, Paper SPE 24584 presented at the SPE Annual Technical Conference and Exhibition, Washington, D.C., October 4-7.

Pistre, V., Kinoshita, T., Endo, T., Schilling, K., Pabon, J., Sinha, B., Plona, T., Ikegami, T. and Johnson, D. (2005) A modular wireline sonic tool for measurements of 3D (azimuthal, radial and axial) formation acoustic properties, Proceedings of the 46th SPWLA Annual Logging Symposium held in New Orleans, Louisiana, United States, June 26-29.

Plona, T.J., Kane, M.R., Sinha, B.K., Walsh, J. and Viloria, O. (2000) Using acoustic anisotropy, Proceedings 41th Annual SPWLA Logging Symposium.

Plumb, R.A. (1994) Influence of composition and texture on the failure properties of clastic rocks. Proceedings of the Eurock 1994. Rock Mechanics in Petroleum Engineering, Balkema, Rotterdam, pp. 13-20.

Plumb, R., Papanastasiou, P. and Last, N. (1998) Constraining the state of stress in tectonically active settings, Proceedings of the SPE/ISRM Conference on Rock Mechanics in Petroleum Engineering: Eurok 98, Trondheim, July 8-10.

Plumb, R.A. and Hickman, S.J. (1985) Stress induced borehole elongation: a comparaison between the four arm dipmeter and yhe borehole televiewer in the Auburn geothermal well, *Journal of Geophysical Research*, Vol. 90, N° B7, pp. 5513-5522.

Poland, J. F. and Lofgren, B. E. (1984) Case History No 9.13 San Joaquin Valley, California, USA, in Poland, J.F., ed., Guidebook to studies of land subsidence due to ground-water withdrawal, UNESCO Studies and Reports in Hydrology, N°. 40, pp. 263-277.

Putot, C.J.M., Guesnon J., Perreau P.J. and Constantinescu A. (2000) Quantifying drilling efficiency and disruption: field data vs. theoretical model., *Journal SPE Drilling & Completion,* Vol. 15, N° 2, pp. 118-125.

Putot, C.J.M. and Constantinescu, A. (1997) Modèle de foration fondé sur le couplage des effets d'évacuation des déblais et de coupe de la roche, Symposium Saint-Venant, Editions ENPC, Paris.

Putot, C. (1995) Modèle de foration fondé sur le couplage de la destruction de la roche et d'évacuation des débris, *Oil & Gas Science and Technology – Rev. IFP*, Vol. 50, N° 4, pp. 491-516.

Putot, C.J.M., Perreau P.J. and Constantinescu A. (1998) Field data vs. Theoretical model to quantify drilling efficiency and disruption, Paper SPE 50579 presented at the SPE European Petroleum Conference, The Hague, Netherlands, October 20-22.

Ramos, G.G., Katahara, K.W., Gray, J.D. and Knox, D.J.W. (1994) Sand production in vertical and horizontal wells in a friable sanstone formation, North Sea, Paper SPE 28065 presented at the SPE/ISRM Conference on Rock Mechanics in Petroleum Engineering, Eurock'94, Delft, Netherlands.

Rasolofosaon, P. and Zinszner, B. (2002) Vérification expérimentale de la formule de Gassmann dans les calcaires poreux, *Oil & Gas Science and Technology, Rev. IFP.*, Vol. 57, N° 2, pp. 129-138

Rasolofosaon, P., and Zinszner, B. (2004) Laboratory petroacoustics for seismic monitoring feasibility study, *The Leading Edge*, Vol. 23, N° 3, pp. 252-258.

Ravi, K., Bosma, M. and Gastebled, O. (2002) Safe and economic gas wells through cement design for life of the well, Paper 75700 presented at the SPE Gas Technology Symposium, Calgary, Alberta, Canada, April 30-May 2.

Reinecker, J., Heidbach, O., Tingay, M., Sperner, B. and Müller, B. (2005) The release 2005 of the world stress map (available online at www.world-stress-map.org)

Renard, G., Nauroy, J.-F., Deruyter, Ch., Moulu, J.-C., Sarda, J.-P. and Le Romancer, J.-F (2000) Production froide des huiles visqueuses, *Oil & Gas Science and Technology – Rev. IFP*, Vol. 55, No. 1, pp. 35-66.

Rhett, D.W. and Teufel, L.W. (1992a) Effect of reservoir stress path on compressibility and permeability of sandstones, Paper SPE 24756 presented at the SPE Annual Technical Conference and Exhibition, Washington, D.C., October 4-7.

Rhett, D.W. and Teufel, L.W. (1992b) Failure criteria for high porosity North Sea Chalk. Proceedings of the Fourth North Sea Chalk Symposium, Deauville, France, September 21-23.

Roscoe, K.H. and Burland, J.B. (1968) On the generalised stress-strain behaviour of 'wet clay', *Engineering Plasticity*, Eds Heyman J. and Leckie, F.A., Cambridge University Press, pp. 535-609.

Rousset, G. (1988) Comportement mécanique des argiles profondes – Application au stockage de déchets radioactifs, PhD Thesis, Ecole Nationale des Ponts et Chaussées.

Ruisten, H., Teufel, L.W. and Rhett, D. (1996) Influence of reservoir stress path on deformation and permeability of weakly cemented sandstone reservoirs, Paper SPE 36535 presented at the SPE Annual Technical Conference and Exhibition, Denver, Colorado, October 6-9.

Rutqvist, J., Birkholzer, J.T. and Chin-Fu Tsang, (2007) Coupled reservoir – geomechanical analysis of the potential for tensile and shear failure associated with CO_2 injection in multilayered reservoir – caprock systems, *International Journal of Rock Mechanics & Mining Sciences,* Vol. 45, pp. 132-143.

Santarelli, F.J., Dardeau, C. and Zurdo, C. (1992) Drilling through highly fractured formations: a problem, a model, and a cure, Paper SPE 24592 presented at the SPE Annual Technical Conference and Exhibition, Washington, D.C., October 4-7.

Santarelli, F.J., Tronvoll J.T., M. Svennekjaer, Skeie H., Henriksen H. and Bratli R.K. (1998) Reservoir stress-path: The depletion and the rebound, Paper SPE 47350 presented at the SPE/ISRM Conference on Rock Mechanics in Petroleum Engineering, Trondheim, Norway, July 8-10.

Sarda, J-P. (1984) Fracturation hydraulique, Rapport IFP 31911.

Sargand, S. M. and Hazen, G. A. (1987) Deformation behaviour of shales, *International Journal Rock Mechanics Mining Sciences*, Vol. 24, No.6, pp. 365-370.

Schofield and Wroth (1968) Critical state soil mechanics, McGraw Hill, London.

Scholz, C. H. (1990) The mechanics of earthquakes and faulting, Cambridge University Press.

Schutjens, P.M.T.M. and de Ruig, H. (1996) The influence of stress path and permeability of an overpressurised reservoir sandstone: some experimental data, *Physics and Chemistry of the Earth*, Vol. 22, N° 1-2, pp. 97-103.

Schutjens, P.M.T.M., Blanton T.L., Martin J.W., Lehr, B.C. and Baaijens, M.N. (1998) Depletion-induced compaction of an overpressurised reservoir sandstone: an experimental approach. Paper SPE 47277 presented at the SPE/ISRM Rock Mechanics in Petroleum Engineering, Trondheim, Norway, July 8-10.

Seidel, F. A. and Greene, T. G. (1985) Use of expanding cement improves bonding and aids in eliminating annular gas migration in Hobbs Grayburg-San Andres wells, Paper SPE 14434 presented at the 60th Annual Technical Conference and Exhibition held in Las Vegas, NV, Sept. 22-25.

Servant, G., Nauroy, J-F. and Bemer, E. (2007) Coupled geomechanical transport modelling of sand production in petroleum reservoirs, Proceedings of the 2nd International Conference on Porous Media and its Applications in Science and Engineering, ICPM2, Kauai, Hawaii, USA.

Settari, A. and Mourits, F.M. (1994) Coupling of geomechanics and reservoir simulation models, *In Computer Methods and Advances in Geomechanics*, Siriwardane and Zaman Editors, Balkema, Rotterdam, pp. 2151-2158.

Settari, A. and Mourits, F.M. (1998) A coupled reservoir and geomechanical simulation system, Paper SPE 50939, *SPE Journal*, September, pp. 219-226.

Settari, A. and Walters, D.A. (2001) Advances in coupled geomechanical and reservoir modelling with applications to reservoir compaction, Paper SPE 74142, *SPE Journal*, Vol. 6, N° 3, pp. 334-342.

Settari, A. (2002) Reservoir compaction, Paper SPE 76805, *Journal of Petroleum Technology*, Vol. 54, N° 8, pp. 62-69.

Shao, J.F. and Marchina, P. (2002) A damage mechanics approach for the modelling of sand production in heavy oil reservoirs, Paper SPE 78167 presented at the SPE/ISRM Rock Mechanics Conference, Irving, Texas, October 20-23.

Sinha, B.K., Kane, M.R. and Frignet, B. (2000) Dipole dispersion crossover and sonic logs in a limestone reservoir, *Geophysics*, Vol. 65, N° 2, pp. 390-407.

Smith, R.C., Beirute, R. M. and Holman, G. B. (1984) Improved method of setting successful cement plugs, paper SPE 11415, *Journal of Petroleum Technology*, Vol. 36, N° 11, pp. 1897-1904.

Steiger, P. S. and Leung, P. K. (1992) Quantitative determination of the mechanical properties of shales. Paper SPE 18024, *SPE Drilling Engineering*, Vol. 7, N° 3, pp. 181-185.

Stone, T., Bowen, G., Papanastasiou, P. and Fuller, J. (2000) Fully coupled geomechanics in a commercial reservoir simulator, Paper SPE 65107 presented at the SPE European Petroleum Conference, Paris, October 24-25.

Sulak, R.M. and Danielsen, J. (1989) Reservoir aspects of Ekofisk subsidence, Paper SPE 17852, *Journal of Petroleum Technology*, Vol. 41, N° 7, pp. 709-716.

Sylte, J.E., Thomas, L.K., Rhett, D.W., Bruning, D.D. and Nagel, N.B. (1999) Water induced compaction in the Ekofisk Field, Paper SPE 56426 presented at the SPE Annual Technical Conference and Exhibition held in Houston, Texas, October 3-6.

Theron, A., De Langre, E. and Putot, C. (2001) The effect of dynamical parameters on precession in rotary drilling, *Journal of Energy Resources Technology*, Vol. 123, N° 3, pp. 181-186.

Tixier, M.P., Loveless, G.W. and Anderson, R.A. (1975) Estimation of formation strength from the mechanical properties log, Paper SPE 4532-PA., *Journal of Petroleum Technology*, Vol. 27, N° 3, pp. 283-293.

Tortike, W.S. and Farouq, A. (1993) Reservoir simulation integrated with geomechanics. *Journal of Canadian Petroleum Technology*, Vol. 32, N° 5, pp. 28-37.

Tremblay, B., Sedgwick, G. and Forshner, K. (1996) Imaging of sand production in a horizontal pack by X-ray computed tomography, Paper SPE 30248-PA, *SPE Formation Evaluation,* Vol. 11, N° 2, pp. 94-98.

Tremblay, B., Sedgwick, G. and Forshner, K. (1996) Simulation of cold production in heavy oil reservoirs: wormhole dynamics, Paper SPE 35387 presented at the SPE/DOE 10th Symposium on Improved Oil Recovery held in Tulsa, Oklahoma, April 21-24.

Tremblay, B., Oldakowski, K. and Settari, A. (1997) Geomechanical properties of oil sands at low effective stress, Paper 97-07 presented at the 48th Annual Technical Meeting of the Petroleum Society in Calgary, Alberta, Canada.

Tremblay, B., Sedgwick, G. and Forshner, K. (1998) Modelling of sand production from wells on primary recovery, *Journal of Canadian Petroleum Technology*, Vol. 37, N° 3.

Tremblay, B., Sedgwick, G., Vu, D., Lillico, D., Jossy, E., Yuan, J-Y, Babchin, A. and Sawatzky, R. (1998) Cold production in heavy oil reservoirs, Poster presented at the 15th World Petroleum Congress, Beijing, China.

Vajdova, V., Baud, P. and Wong, T-F. (2004) Compaction, dilatancy and failure in porous carbonate rocks, *Journal of Geophysical Research*, Vol. 109, N° B5, B05204.

Van Hasselt, J.P. (1992) Reservoir compaction and surface subsidence resulting from oil and gas-production: a review of theoretical and experimental research approaches, *Geologie En Mijnbouw*, Vol. 71, N° 2, pp. 107-118.

Van Oort, E., Nicholson, J. and D'Agostino, J. (2001) Integrated borehole stability studies: key to drilling at the technical limit and trouble cost reduction, Paper SPE 67763 presented at the SPE/IADC Drilling Conference, Amsterdam, Netherlands, February 27 -March 1.

Vardoulakis, I., Stavropoulou, M. and Papanastasiou, P. (1996) Hydro-mechanical aspects of the sand production problem, *Transport in porous media*, Vol. 22, N° 2, pp 225-244.

Vaziri, H., Barree, B., Xiao, Y., Palmer, I. and Kutas, M. (2002) What is the magic of water in producing sand?, Paper SPE 77683 presented at the SPE Annual Technical Conference and Exhibition, San Antonio, Texas, September 29 -October 2.

Vaziri, H. (2004) Integrated soil and rock mechanics: a major step towards consistent and improved sand production prediction and management, Paper 04-461 presented at the 6th North America Rock Mechanics Symposium (NARMS), Houston, Texas, June 5-9.

Veeken, C.A. M., Davies, D. R., Kenter, C. J. and Kooijman, A. P. (1991) Sand production prediction review: developing an integrated approach, Proceedings 66th Annual Technical Conference of the SPE, Dallas, Paper SPE 22792 presented at the SPE Annual Technical Conference and Exhibition, Dallas, Texas, October 6-9.

Vernik L., Bruno M. and Bovberg, C. (1993) Empiricial relations between compressive strength and porosity of siliciclastic rocks. *International. Journal of Rock Mechanics and Mining Science,* Vol. 30, No.7, pp. 677-780.

Vidal, S., Longuemare, P. and Huguet, F. (2000) Integrating geomechanics and geophysics for reservoir seismic monitoring, feasibility studies, Paper SPE 65157 presented at the SPE European Petroleum Conference, Paris, October 24-25.

Vidal, S., Huguet, F. and Mechler, P. (2002) Characterising reservoir parameters by integrating seismic monitoring and geomechanics, *The Leading Edge*, Vol. 21, pp. 295-301.

Vidal-Gilbert, S., Huguet, F., Assouline, L. and Longuemare, P. (2005) Hydromechanical modelling of reservoir behaviour during underground gas storage exploitation, Proceeding of the 68th European Association of Geoscientists and Engineers Conference & Exhibition, Madrid, Spain, June 13-16.

Vidal-Gilbert, S., Nauroy, J-F. and Brosse, E., (2009) 3D geomechanical modelling for CO_2 geologic storage in the Dogger carbonates of the Paris Basin, *International Journal of Greenhouse Gas Control*, Vol. 3, N° 3, pp. 288-299.

Vincké, O. (1994) An estimation of bulk moduli of sandstones as a function of confining pressure using their petrographic and petrophysic description, Paper SPE/ISRM 28038 presented at the SPE Conference on Rock Mechanics in Petroleum Engineering, Delft, Netherlands, August 29-31.

Vincké, O., Boutéca, M. J., Piau, J. M. and Fourmaintraux, D. (1998) Study of the effective stress at failure, Proceedings of the Biot Conference on Poromechanics, Louvain-La-Neuve, Belgique, September 14-16, Thimus *et al* Editions, Balkema, Rotterdam, pp. 635-639.

Wang, Z. (2000) Dynamic versus static elastic properties of reservoir rocks, in seismic and acoustic velocities in reservoir rocks, SEG Geophysics Reprint Series, No 19.

Ward, C.D., Coghil, K. and Broussard, M.D. (1994) The application of petrophysical data to improve pore and fracture pressure determination in North Sea Central Graben HPHT wells, Paper SPE 28297 presented at the SPE Annual Technical Conference and Exhibition, New Orleans, Louisiana, September 25-28.

Wardlaw, H. W. R (1971) Optimization of rotary drilling parameters, PhD thesis, Texas University, Austin.

Warpinski, N.R., Moschovidis, Z.A., Parker, C.D. and Abou-Sayed, I.S. (1994) Comparison study of hydraulic fracturing models test case: GRI staged field experiment N° 3, Paper SPE 275890, *SPE Production & Facilities*, Vol. 9, N° 1, pp. 7-16.

Warren, T.M. (1984) Factors affecting torque for a roller cone bit, Paper SPE 11991-P.A., *Journal of Petroleum Technology*, Vol. 136, N° 9, pp. 1500-1508.

Warren, T.M. and Smith, M.B. (1985) Bottomhole stress factors affecting drilling rate at depth, Paper SPE 13381, *Journal of Petroleum Technology*, August, pp. 1523-1535.

Williamson, M.D., Murray, S. J., Atkins, W.S., Hamilton, T.A.P. and Copland, M.A. (1997) A review of Zechstein drilling issues, Paper SPE 51182, *SPE Drilling & Completion*, Vol. 13, N° 3, pp. 174-181.

Wong T-F., David C. and Zhu, W. (1997) The transition from brittle to cataclastic flow: mechanical deformation, *Journal of Geophysical Research*, Vol. 102, N° B2, pp. 3009-3025.

Wright, C.A., Weijers, L., Davis, E.J. and Mayerhofer, M. (1999) Understanding hydraulic fracture growth: tricky but not hopeless, Paper SPE 56724 presented at the SPE Annual Technical Conference and Exhibition, Houston, Texas, October 3-6.

Wriput, D. and Zoback, M.D. (2000) Fault reactivation and fluid flow along a previously dormant normal fault in the northern North Sea, *Geology*, Vol. 28, pp. 595-598.

Xu, W., Le Calvez, J. and Thiercelin, M. (2009) Characterisation of hydraulically-induced fracture network using treatment and microseismic data in a tight-gas formation: a geomechanical approach, Paper SPE 125237 presented at the SPE Tight Gas Completions Conference held in San Antonio, Texas, June 15-17.

Yalamas, T., Nauroy, J.-F., Bemer, E., Dormieux, L. and Garnier, D. (2004) Sand erosion in cold heavy oil production, Paper SPE 86949 presented at SPE International Thermal Operations and Heavy Oil Symposium and Western Regional Meeting, Bakersfield, California, March 16-18.

Yale, D. P., Nieto, J.A. and Austin, S. P. (1995) The effect of cementation on the static and dynamic mechanical properties of the Rotliegendes sandstone, In: Daemen, J.J.K. & Schulz, R. A. (ed.) Proceedings of the 35th U.S. Symposium on Rock Mechanics, Balkema, Rotterdam, pp. 169-175.

Yale, D.P. and Crawford, B. (1998) Plasticity and Permeability in Carbonates: Dependance on Stress Path and Porosity, Paper SPE 47582 presented at the SPE/ISRM Conference on Rock Mechanics in Petroleum Engineering, Trondheim, Norway, July 8-10.

Yamamoto, K., Gutierrez, M., T. Shimamoto and Maezumi (2000) Verification of a 3D hydraulic fracturing model against a field case, Paper SPE 59373 presented at the SPE/DOE Improved Oil Recovery Symposium, Tulsa, Oklahoma, April 3-5.

Yudovich, A., Chin, L.Y. and Morgan, D.R. (1989) Casing deformation in Ekofisk, Paper SPE 17856, *Journal of Petroleum Technology*, Vol. 41, N° 7, pp. 729-734.

Zhang, W., Youn, S. and Doan, Q. (2005) Understanding reservoir architectures and steam chamber growth at Christina lake, Alberta, by using 4D seismic and crosswell seismic imaging, Paper SPE 97808 presented at the SPE/PS-CIM/CHOA International Thermal Operations and Heavy Oil Symposium, Calgary, Alberta, Canada, November 1-3.

Zhu, W. and Wong, T.F. (1997). The transition from brittle to cataclastic flow: permeability evolution, *Journal of Geophysical Research*, Vol. 102, B2, Feb. 10, pp. 3027-3041.

Zimmerman, R. W. (1991) Compressibility of sandstones, Elsevier Science Publishers B.V., Amsterdam.

Zinszner, B. and Pellerin, F-M. (2007) A Geoscientist's guide to petrophysics, Editions Technip, Paris

Zoback, M.D., Barto, C.A., Brudy, M.O., Castillo, D.A., Finkbeiner, T., Grollimund, B.R., Moos, D.B., Peska, P., Ward, C.D. and Wiprut, D.J. (2003) Determination of stress orientation and magnitude in deep wells, *International Journal of Rock Mechanics & Mining Sciences,* Vol. 40, pp. 1049-1076.

Index